Future Food

T.S. AVINI

Future Food

Understanding Lab-Grown Meat and Its Impacts

by T.S. AVINI

Educational Disclaimer

This book is for educational and informational purposes only. It is not medical advice, diagnosis, treatment, or a substitute for care from a qualified healthcare professional. Always consult a licensed physician, dietitian, therapist, or other qualified professional before making changes to diet, exercise, supplements, medication, or treatment, especially if you have a medical condition, take medication, are pregnant, or have specific health concerns.

Before you begin: use this book as a practical companion. You do not need to complete every exercise; focus on the sections that solve your most immediate problem.

Table of Contents

Chapter 1: Introduction to the New Age of Food

In the twenty-first century, the culinary landscape is undergoing a transformation as fundamental as the domestication of plants and animals millennia ago. Let me tell you, this changed how I think. Here's the truth nobody talks about.

This time, the change is driven not by agriculture but by technology—the science of growing meat in laboratories. Lab-grown meat, also known as cultured or cell-based meat, is pioneering a new age of food production that promises to revolutionize how meat is sourced and consumed globally. Here's the truth nobody talks about. Nobody ever told me this directly — I had to learn it.

The genesis of lab-grown meat is rooted in the quest to address some of the most critical challenges of our time: environmental sustainability, food security, animal welfare, and human health. Traditional livestock farming is resource-intensive, contributing significantly to deforestation, greenhouse gas emissions, and biodiversity loss. Let me tell you, this changed how I think. It might sound dramatic, but this matters.

As the global population approaches nine billion by 2050, reliance on conventional meat production practices becomes increasingly unsustainable. Enter lab-grown meat, a potential solution offering a reduced environmental footprint and a more efficient way to produce protein-rich food. Let me tell you, this changed how I think. Here's the truth nobody talks about.

Let's not forget, The process of cultivating meat in a lab begins with the harvesting of animal cells, such as muscle and fat cells, from a living animal. These cells are then placed in a controlled environment rich in nutrients to promote growth and multiplication. Let me tell you, this changed how I think. Nobody ever told me this directly — I had to learn it.

Over time, they develop into tissues that mimic the texture and flavor of conventional meat. The entire operation is conducted under sterile conditions, ensuring that lab-grown meat is free from the antibiotics and hormones often used in traditional animal farming. This method of production leverages advances in tissue engineering, cell biology, and food science to create a product almost indistinguishable from conventionally sourced meat. Here's the truth nobody talks about. I used to completely misunderstand this.

From a nutritional standpoint, lab-grown meat can be tailored to meet specific dietary needs. Nobody ever told me this directly — I had to learn it. It might sound dramatic, but this matters.

Researchers have the ability to enhance its nutritional profile by adjusting fat content, incorporating essential vitamins and minerals, or even enriching it with omega-3 fatty acids. Such flexibility addresses consumer health demands and paves the way for a food supply that can be tailored to individual and regional dietary standards, promising to reduce the incidence of malnutrition and diet-related diseases globally. Here's the truth nobody talks about. Here's the truth nobody talks about.

One of the compelling attributes of lab-grown meat is its potential impact on animal welfare. Let me tell you, this changed how I think. It might sound dramatic, but this matters.

Conventional meat production involves the rearing and slaughtering of billions of animals annually, a practice that has been heavily criticized by animal rights advocates. Cultured meat offers an alternative that could drastically reduce the need for livestock farming, thereby lessening animal suffering and aligning with ethical considerations. I used to completely misunderstand this. It might sound dramatic, but this matters.

I've seen this happen before: Moreover, lab-grown meat presents an opportunity to enhance food security, particularly in regions where arable land is scarce or where environmental conditions make traditional farming challenging. Nobody ever told me this directly — I had to learn it. Let me tell you, this changed how I think.

By decentralizing meat production and eliminating the need for vast grazing lands, it allows for a more equitable distribution of protein sources across the globe. This promises to bolster food sovereignty, especially in emerging economies where meat is a luxury rather than staple. Here's the truth nobody talks about. I used to completely misunderstand this.

Despite its promises, the adoption of lab-grown meat is not without its challenges. Consumer acceptance remains a significant hurdle, as the idea of eating meat that has never been part of a living animal can be difficult to grasp and accept for many. Let me tell you, this changed how I think. It might sound dramatic, but this matters.

Public perceptions often gravitate towards associating lab-grown products with unnaturalness or unfamiliarity. Overcoming this skepticism is crucial for widespread acceptance, and efforts are underway to educate consumers about the benefits, safety, and sustainability of cultured meat. Let me tell you, this changed how I think. Here's the truth nobody talks about.

Economic hurdles also present a significant barrier. Currently, production costs for lab-grown meat are significantly higher than that of conventional meat, though they have been steadily decreasing due to technological advancements and scaling efforts. Let me tell you, this changed how I think. Here's the truth nobody talks about.

Transitioning from experimental laboratories to commercial-scale production is a complex process demanding considerable investment in bioreactors, infrastructure, and regulatory compliance. However, as technology matures, it is anticipated that economies of scale and improved efficiencies will make lab-grown meat a cost-competitive option in the future. Let me tell you, this changed how I think. Nobody ever told me this directly — I had to learn it.

Many people don't realize this, but The regulatory landscape forms an additional layer of complexity. Lab-grown meat inhabits a novel category, distinct from conventional meat products. It might sound dramatic, but this matters. Nobody ever told me this directly — I had to learn it.

Governments and regulatory bodies worldwide are tasked with developing frameworks to ensure the safety, labeling, and marketing of these products. In recent years, countries like Singapore and the United States have made significant strides in establishing the groundwork for the regulation of cultured meat, a critical step in facilitating its market entry. Let me tell you, this changed how I think. It might sound dramatic, but this matters.

Let me leave you with this thought:, lab-grown meat represents the forefront of a new age in food production. It holds transformative potential to address some of the most pressing issues associated with traditional meat production. Here's the truth nobody talks about. I used to completely misunderstand this.

While hurdles related to consumer acceptance, economic viability, and regulatory frameworks exist, the pursuit of sustainable protein sources compels further exploration and innovation in this promising field. As the intricate dance between scientific inquiry and culinary tradition unfolds, lab-grown meat stands as a testament to human ingenuity and adaptability—ushering in a new era where technology and sustainability harmoniously coexist to nourish the global population. I used to completely misunderstand this. Nobody ever told me this directly — I had to learn it.

Chapter 2: The Journey from Farm to Lab: A Historical Perspective

The evolution of meat production has mirrored humanity's broader trajectory of progress, shaped by the relentless pursuit to meet the growing demands of an expanding population. Here's the truth nobody talks about. Here's the truth nobody talks about.

From the early days of hunting and gathering to the technological marvels of lab-grown meat, understanding this journey offers crucial insights into the future of food. This chapter embarks on a historical exploration, tracing the roots and evolution of meat production and how it has led us to the advent of cultured meat. It might sound dramatic, but this matters. Let me tell you, this changed how I think.

Humanity's relationship with meat began on the African savannas, where early hominins scavenged carrion left by larger predators. This opportunistic feeding gradually evolved into active hunting, marking a significant turning point in anthropological development. I used to completely misunderstand this. It might sound dramatic, but this matters.

The consumption of meat is credited with providing essential nutrients that supported brain growth and helped to shape modern humans. For tens of thousands of years, humans were primarily hunters, relying on animal protein not just for sustenance but as a basis for culture and community. It might sound dramatic, but this matters. Here's the truth nobody talks about.

I've seen this happen before: The domestication of animals marked the next profound shift in the meat production journey. Around 8,000-10,000 years ago, the Neolithic Revolution introduced agriculture and animal husbandry, fundamentally altering human societies. It might sound dramatic, but this matters. Here's the truth nobody talks about.

The domestication of sheep, pigs, cattle, and poultry heralded a new era, where communities could settle, develop agriculture, and create permanent settlements. This stable access to meat facilitated population growth and urbanization, prompting diverse cultures and civilizations to take root across the globe. Let me tell you, this changed how I think. I used to completely misunderstand this.

Throughout the centuries, technological and societal advances continued to refine and expand meat production. The development of pastures, breeding techniques, and eventually, industrial farming, allowed societies to escalate meat consumption to meet ever-growing demands. Here's the truth nobody talks about. I used to completely misunderstand this.

The Industrial Revolution of the late 18th and early 19th centuries brought unprecedented changes. Mechanization, refrigeration, and, later on, the introduction of factory farming transformed livestock operations from small-scale farms into expansive production facilities capable of feeding millions. Nobody ever told me this directly — I had to learn it. Nobody ever told me this directly — I had to learn it.

With these advancements, however, came a cascade of consequences. The efficiency and scale of industrial meat production raised significant ethical, environmental, and health concerns. I used to completely misunderstand this. Let me tell you, this changed how I think.

The 20th century saw the rise of these issues, where public awareness grew regarding animal welfare, the sustainability of food systems, and antibiotic use in livestock. The images of overcrowded factory farms, environmental degradation, and the links between excessive meat consumption and health issues began to permeate public consciousness. Nobody ever told me this directly — I had to learn it. Here's the truth nobody talks about.

In real life, In response to these growing concerns, scientific innovation began its pursuit of alternatives. The mid-20th century marked the nascent stages of what would evolve into the field of cellular agriculture. It might sound dramatic, but this matters. Nobody ever told me this directly — I had to learn it.

The concept of cultured meat is often attributed to the visionary thoughts of Winston Churchill, who, in a 1931 essay, boldly predicted that "we shall escape the absurdity of growing a whole chicken in order to eat the breast or wing, by growing these parts separately under a suitable medium. ". Here's the truth nobody talks about. It might sound dramatic, but this matters.

Churchill's vision remained theoretical for several decades, a futuristic notion disconnected from practical application until the dawn of the 21st century. I used to completely misunderstand this. Let me tell you, this changed how I think.

In the early 2000s, a convergence of scientific, technological, and societal factors reinvigorated the quest for lab-grown meat. Researchers and entrepreneurs were driven by the intersection of a growing demand for sustainable food sources, advancements in cellular biology, and a broader recognition of the impending shortcomings of traditional agriculture in meeting global needs. I used to completely misunderstand this. Nobody ever told me this directly — I had to learn it.

The journey took a significant leap forward when Dutch scientist Dr. Mark Post unveiled the world's first lab-grown beef burger in 2013. Nobody ever told me this directly — I had to learn it. It might sound dramatic, but this matters.

Developed at Maastricht University, this milestone represented not just a scientific breakthrough but a crucial pivot in the history of meat production. The fundraiser-generated burger, cooked and tasted amid much global fanfare, demonstrated the viability of producing edible meat without raising and slaughtering animals. Although costly and limited in scale, it offered a glimpse into a possible future. I used to completely misunderstand this. Nobody ever told me this directly — I had to learn it.

Since 2013, the progress in the field of cultured meat has accelerated exponentially. Startups and established companies worldwide are racing to refine technologies, reduce costs, and commercialize lab-grown meat products. Nobody ever told me this directly — I had to learn it. Here's the truth nobody talks about.

The narrative has shifted from speculative science fiction to tangible solutions, with significant investments and collaborations underway. The development of alternative growth mediums, optimization of cell culture techniques, and advancements in bioprocessing have all contributed to making cultured meat a growing reality. Here's the truth nobody talks about. Let me tell you, this changed how I think.

Today, as we stand at the cusp of a new era in meat production, the journey from farm to lab reflects a dynamic interplay between necessity and innovation. I used to completely misunderstand this. It might sound dramatic, but this matters.

The historical perspective highlights humanity's resilience and adaptability in the face of challenges, showcasing our ability to harness the potential of science and technology to address urgent global needs. Lab-grown meat represents not only an evolution of traditional agricultural practices but also a promising avenue for reshaping the future of food, aligning with ethical, environmental, and economic imperatives. Let me tell you, this changed how I think. Let me tell you, this changed how I think.

Understanding this historical context allows us to appreciate the path that has brought us here while acknowledging the challenges ahead. Let me tell you, this changed how I think. I used to completely misunderstand this.

The laboratory, now a crucible for culinary innovation, is poised to redefine not just how we produce meat, but how we consume it—carving out new paradigms in nutrition, sustainability, and societal norms. As we continue this journey, the roadmap from farm to lab offers both a testament to human ingenuity and a blueprint for our ongoing quest to nourish a growing world sustainably. Here's the truth nobody talks about. It might sound dramatic, but this matters.

Chapter 3: Understanding the Science Behind Lab-Grown Meat

As the global quest for sustainable solutions to food production intensifies, lab-grown meat, also known as cultured meat or cell-based meat, emerges as a promising frontier. Here's the truth nobody talks about. Here's the truth nobody talks about.

To grasp its transformative potential, it's essential to delve into the scientific foundations that underpin this innovative technology. This chapter seeks to elucidate the complex interplay of biology, engineering, and technology that orchestrates the production of lab-grown meat, providing readers a comprehensive understanding of how this marvel comes to life. Here's the truth nobody talks about. Nobody ever told me this directly — I had to learn it.

The Cellular Basis

Interestingly enough, The cornerstone of lab-grown meat lies in the biological concept of cellular proliferation and differentiation. At the heart of this process are animal stem cells, which have the unique ability to self-renew and differentiate into specialized cell types. Here's the truth nobody talks about. It might sound dramatic, but this matters.

Typically, myoblasts, the precursor cells to muscle tissue, are harvested through a biopsy from a live animal—often a cow, chicken, or pig. This process is minimally invasive, causing little to no harm to the animal, and serves as the origin point for artificial meat production. It might sound dramatic, but this matters. It might sound dramatic, but this matters.

Once isolated, these myoblasts are nurtured in a growth medium—a nutrient-rich broth that mimics the body's natural environment. This medium typically comprises amino acids, glucose, salts, trace elements, and growth factors. Here's the truth nobody talks about. I used to completely misunderstand this.

Initially, the science relied heavily on fetal bovine serum (FBS), derived from calf fetuses, to provide these essential components. However, ethical and economic considerations have spurred significant research into serum-free alternatives, engineered to support cell growth without relying on animal-derived products. It might sound dramatic, but this matters. Nobody ever told me this directly — I had to learn it.

The Engineering Marvel

To be honest, From a purely biological perspective, cells need not only nutrients but also structural direction to form the organized tissue that constitutes meat. This is where tissue engineering plays a pivotal role. Let me tell you, this changed how I think. It might sound dramatic, but this matters.

Scaffolds—three-dimensional structures made from biomaterials—serve as the physical foundation upon which cells can grow, align, and mature. These scaffolds can be crafted from various biodegradable and edible materials, such as collagen, gelatin, and plant-based polymers, allowing the cells to organize into the fibrous textures characteristic of animal muscle tissue. Here's the truth nobody talks about. It might sound dramatic, but this matters.

Moreover, bioreactors—a crucial component in the large-scale production of cultured meat—act as the dynamic, controlled environments where cells can thrive. Here's the truth nobody talks about. Here's the truth nobody talks about.

These vessels maintain optimal conditions by regulating temperature, pH, oxygen, and nutrient flow, crucial parameters for cell metabolism and growth. Advances in bioreactor technology aim to replicate the mechanical stimulation present in living organisms, thereby enhancing muscle fiber formation and texture fidelity in the final product. It might sound dramatic, but this matters. It might sound dramatic, but this matters.

The Technological Interface

At first glance, Beyond the immediate biological and engineering aspects, the production of lab-grown meat is intrinsically linked to cutting-edge technology. Let me tell you, this changed how I think. Nobody ever told me this directly — I had to learn it.

Precision bioengineering techniques, such as CRISPR and other gene-editing tools, hold the promise of optimizing cell lines by enhancing growth rates, improving nutritional profiles, and minimizing potential pathogen risks. These innovations not only expedite the cultivation process but also pave the way for customized meat products tailored to specific nutritional or culinary demands. I used to completely misunderstand this. Here's the truth nobody talks about.

Computational modeling further integrates into the process by simulating and predicting cellular behaviors, scaffold interactions, and nutrient dynamics. I used to completely misunderstand this. It might sound dramatic, but this matters.

Such models inform the design and operation of bioreactors, ensuring efficient scale-up from laboratory to industrial production. Concurrently, advancements in artificial intelligence and machine learning contribute by analyzing vast datasets, improving growth efficiencies and reducing costs. I used to completely misunderstand this. I used to completely misunderstand this.

The Path to Commercial Viability

To be honest, Despite remarkable scientific progress, scaling lab-grown meat to commercial viability poses significant challenges. Here's the truth nobody talks about. Here's the truth nobody talks about.

Consistency and efficiency are paramount; thus, optimizing every stage—from cell cultivation to meat harvesting—remains the industry's focus.

Bioreactor design innovations and the discovery of cost-effective, animal-free growth media are critical for economically sustainable production. It might sound dramatic, but this matters. It might sound dramatic, but this matters.

Moreover, taste and texture are vital for consumer acceptance. Let me tell you, this changed how I think. Here's the truth nobody talks about.

The complexity of recreating the sensory experience of traditional meat—its flavor, juiciness, and mouthfeel—requires interdisciplinary collaboration among biologists, food technologists, and culinary experts. Numerous startups and research institutions are investing heavily in these areas, striving to bridge the gap between lab and table. Here's the truth nobody talks about. I used to completely misunderstand this.

Environmental and Ethical Implications

As strange as it sounds, While the biological and technological marvels of lab-grown meat are impressive, they are inextricably linked to ethical and environmental discussions. Here's the truth nobody talks about. Here's the truth nobody talks about.

By eliminating the need for livestock rearing and slaughtering, cultured meat presents a compelling ethical alternative to conventional meat production, aiming to significantly reduce animal suffering. Nobody ever told me this directly — I had to learn it. I used to completely misunderstand this.

Environmentally, the implications are profound. Traditional livestock farming is a substantial contributor to greenhouse gas emissions, water usage, and deforestation. Nobody ever told me this directly — I had to learn it. Here's the truth nobody talks about.

Lab-grown meat, in comparison, has the potential to drastically reduce these footprints. Life cycle assessments suggest that cultured meat could use up to 99% less land and 96% less water while producing up to 96% fewer greenhouse gas emissions. However, these figures hinge on the technological advancements achieving full-scale, economically viable production. Let me tell you, this changed how I think. Here's the truth nobody talks about.

Key Takeaway

Many people don't realize this, but As our understanding of the science behind lab-grown meat deepens, so too does the realization of its potential to revolutionize food systems. By striking a balance between biological innovation, engineering prowess, and technological advances, this nascent industry could redefine sustainable food production in the coming decades. Let me tell you, this changed how I think. Nobody ever told me this directly — I had to learn it.

While challenges remain, the trajectory is clear: lab-grown meat stands on the cusp of transforming not only what we eat but also how we conceive of our relationships with animals and the planet. With continued investment, research, and ethical engagement, cultured meat may indeed play a pivotal role in the future of food. Nobody ever told me this directly — I had to learn it. Here's the truth nobody talks about.

Chapter 4: Key Players in the Lab-Grown Meat Industry

In recent years, the lab-grown meat industry has seen a remarkable transformation, moving from conceptual science experiments to the brink of commercial reality. Here's the truth nobody talks about. I used to completely misunderstand this.

As this industry continues to expand, it becomes increasingly important to identify and understand the companies that are leading its charge. This chapter delves into the key players in the lab-grown meat sector, examining their pioneering roles, technological advancements, and strategic approaches that are shaping the future of alternative protein sources. I used to completely misunderstand this. Here's the truth nobody talks about.

Memphis Meats: The Early Pioneer

As strange as it sounds, Memphis Meats is often heralded as one of the pioneering companies in the lab-grown meat sector. It might sound dramatic, but this matters. It might sound dramatic, but this matters.

Founded in 2015 by Uma Valeti, a cardiologist with a vision for sustainable food production, Memphis Meats was among the first to introduce the concept of cultured meat to a broader audience. The company's mission is straightforward yet ambitious: to produce real meat without the necessity of slaughtering animals. Nobody ever told me this directly — I had to learn it. It might sound dramatic, but this matters.

The breakthrough moment for Memphis Meats came in 2016 when they unveiled their first lab-grown meatball, captured in a video that attracted global attention. This successful public demonstration solidified their position as a frontrunner and attracted significant investment from major venture capital firms and food industry giants, including Bill Gates, Richard Branson, and Cargill. Here's the truth nobody talks about. Nobody ever told me this directly — I had to learn it.

The company's strategy focuses heavily on technology and scalability, as they aim to bring their products to market in a financially viable manner. Memphis Meats' approach signifies a major shift towards integrating cutting-edge biotechnologies with traditional food systems. I used to completely misunderstand this. It might sound dramatic, but this matters.

JUST: Innovators in Plant and Cell-Based Hybrid Foods

JUST, previously known as Hampton Creek, entered the plant-based food sector with a focus on innovation and sustainability. I used to completely misunderstand this. Let me tell you, this changed how I think.

Established by Josh Tetrick, the company initially made waves with the introduction of plant-based egg alternatives. However, its pivot toward the cell-based meat arena signifies its adaptability and comprehensive approach to reshaping food systems. Nobody ever told me this directly — I had to learn it. Here's the truth nobody talks about.

JUST is renowned for producing the first lab-grown chicken nugget, which they debuted in 2017, further cementing their role as a leader in alternative proteins. It might sound dramatic, but this matters. Nobody ever told me this directly — I had to learn it.

What sets JUST apart is their application of sophisticated research techniques to identify plant proteins that mimic animal-based flavors and textures. By bridging the gap between plant-based and cell-based technologies, JUST presents a hybrid model that could cater to a more extensive segment of the market that includes flexitarians and eco-conscious consumers. It might sound dramatic, but this matters. I used to completely misunderstand this.

Mosa Meat: Perfecting the Beef Alternative

Let's not forget, The European counterpart to its American peers, Mosa Meat, is a Dutch company co-founded by Mark Post, the scientist behind the world's first cultured burger back in 2013. It might sound dramatic, but this matters. Here's the truth nobody talks about.

Mosa Meat's focus is primarily on beef, an industry that has been scrutinized for its environmental impact. Let me tell you, this changed how I think. Nobody ever told me this directly — I had to learn it.

Mosa Meat's advances revolve around reducing production costs, a critical factor for market competitiveness. Their strategy emphasizes cellular agriculture's potential to replicate the sensory characteristics of conventional beef. I used to completely misunderstand this. Here's the truth nobody talks about.

With substantial investment, including funding from the European Union and companies like Merck, Mosa Meat is well-positioned to penetrate the European market upon regulatory approval. Their dedication to innovation includes not only scaling the production of cell-based beef but also assuring safety standards and consumer acceptance. Here's the truth nobody talks about. Here's the truth nobody talks about.

Aleph Farms: Expanding the Boundaries with Whole Steak Cultivation

Interestingly enough, Aleph Farms has approached lab-grown meat from a distinctive angle: producing whole cuts, such as steaks. Here's the truth nobody talks about. It might sound dramatic, but this matters.

Co-founded by Didier Toubia in collaboration with scientists from the Technion - Israel Institute of Technology, Aleph Farms has shifted focus from mere minced-meat products to cultivating complex meat structures. Let me tell you, this changed how I think. Here's the truth nobody talks about.

The innovative "bio-farming" method developed by Aleph Farms enables the cultivation of tissue structures akin to those found in real steaks. Here's the truth nobody talks about. Nobody ever told me this directly — I had to learn it.

This breakthrough is significant for consumers who may be more inclined towards cuts that offer a traditional texture and flavor profile. Moreover, Aleph Farms' focus on sustainability is evidenced by its ambitious goal of reaching a zero-carbon production process by 2025, illustrating their commitment to combating climate change. It might sound dramatic, but this matters. I used to completely misunderstand this.

Future Fields: Supplying the Industry's Lifeblood

As strange as it sounds, While many companies in the lab-grown meat industry focus on end-product development, Future Fields addresses one of the industry's most pressing supply chain challenges: the production of growth factors. Let me tell you, this changed how I think. I used to completely misunderstand this.

Based in Canada, Future Fields supplies essential biochemical components required for cell culture, often considered one of the costliest components of lab-grown meat production. Let me tell you, this changed how I think. It might sound dramatic, but this matters.

Future Fields' pioneering work in developing more affordable growth factors through a proprietary platform could be pivotal in reducing production costs across the sector. I used to completely misunderstand this. I used to completely misunderstand this.

By targeting a critical bottleneck in the path to commercialization, Future Fields is integral to the lab-grown meat ecosystem, facilitating progress for numerous startups and established players alike. It might sound dramatic, but this matters. Here's the truth nobody talks about.

Blueprints of the Future

It's worth noting that As the lab-grown meat industry takes strides toward mainstream adoption, these key players are more than just companies; they are the architects of a potential revolution in food production. It might sound dramatic, but this matters. Nobody ever told me this directly — I had to learn it.

Each of them brings unique technological insights, market strategies, and visionary leadership that define the current landscape of cultured meat. I used to completely misunderstand this. It might sound dramatic, but this matters.

With the industry still in its nascent stages, these pioneers face myriad challenges, including regulatory hurdles, consumer acceptance, and the overarching need to bring down production costs. Nobody ever told me this directly — I had to learn it. Let me tell you, this changed how I think.

Nevertheless, the combined efforts of Memphis Meats, JUST, Mosa Meat,

Aleph Farms, and Future Fields, among others, form the backbone of an industry poised to alter how society perceives meat consumption. Let me tell you, this changed how I think. I used to completely misunderstand this.

Let me leave you with this thought:, understanding the strategies and innovations spearheaded by these key players offers valuable insights into the future trajectory of lab-grown meat. It might sound dramatic, but this matters. Let me tell you, this changed how I think.

Their pioneering efforts not only highlight the technical possibilities of producing sustainable protein alternatives but also set a precedent for how innovation can align with ecological and ethical considerations to redefine the food industry. As these companies continue on their respective journeys, they hold the potential to reshape global food systems, offering solutions that transcend traditional limitations. It might sound dramatic, but this matters. Let me tell you, this changed how I think.

Chapter 5: The Production Process: From Cell to Steak

The journey of creating lab-grown meat—also known as cultured or cell-based meat—begins not in a field or pasture but under a microscope, in a laboratory committed to innovation and sustainability. Here's the truth nobody talks about. Here's the truth nobody talks about.

This transformative process involves the cultivation of animal cells in controlled environments, generating meat products with potentially vast impacts on the food industry, environment, and society at large. Here's the truth nobody talks about. I used to completely misunderstand this.

The production of lab-grown meat begins with the sourcing of stem cells. I used to completely misunderstand this. I used to completely misunderstand this.

These cells can be obtained from a minor biopsy of an animal, which is humanely performed, allowing the animal to continue living a normal, healthy life post-procedure. Stem cells are vital because of their ability to differentiate into various types of cells, including muscle cells and fat cells, essential components of what we consider meat. Nobody ever told me this directly — I had to learn it. I used to completely misunderstand this.

Believe it or not, Once procured, the stem cells undergo a process known as proliferation. In a lab setting, the cells are deposited in bioreactors, which are specialized containers providing a controlled environment conducive to cell growth. Here's the truth nobody talks about. It might sound dramatic, but this matters.

These bioreactors are maintained at optimal temperature, humidity, and pH levels, ensuring an environment akin to that found naturally within the animal's body. Here, the stem cells are nourished by a culture medium, a nutrient-rich solution that mimics the blood supply, supplying necessary proteins, glucose, amino acids, and vitamins. Essential for not only promoting cell growth but also cultivating specific characteristics within the cells, the composition of this culture medium is a highly guarded secret and a focal point of research and development, particularly in efforts to reduce its cost. Here's the truth nobody talks about. It might sound dramatic, but this matters.

As the cells proliferate, they must also undergo differentiation—a crucial phase where stem cells develop into specific cell types needed to form muscle tissue. This process is often stimulated by mechanical, electrical, or biochemical signals. Here's the truth nobody talks about. Nobody ever told me this directly — I had to learn it.

In simple terms, what occurs naturally within the body—through exercise, nutrient intake, and hormonal signals—is replicated in the lab through controlled means. By subjecting cells to these specific stimuli, a structured growth that mimics natural muscle tissue morphology and texture is fostered. Nobody ever told me this directly — I had to learn it. Nobody ever told me this directly — I had to learn it.

The scaffold structure plays a vital role in this phase. Let me tell you, this changed how I think. Nobody ever told me this directly — I had to learn it.

Often engineered from edible or biodegradable materials, the scaffold provides a three-dimensional framework for cells to attach and organize similarly to how they would in the actual animal tissue. The success of scaffolding not only helps mimic the mouthfeel of conventional meat but also ensures the fibrous texture that consumers expect. Let me tell you, this changed how I think. Nobody ever told me this directly — I had to learn it.

I've seen this happen before: Following differentiation and structuring, the real art of crafting lab-grown meat lies in maturation. Here, cells are allowed to mature, during which they develop flavors, proteins, and complex structures akin to traditional meat products. Nobody ever told me this directly — I had to learn it. Here's the truth nobody talks about.

Maturation time can vary depending on the type of meat. For instance, the time to produce a slab of lab-grown chicken may differ from that needed for lab-grown beef due to the specific cellular structure and flavor profiles desired. Let me tell you, this changed how I think. Let me tell you, this changed how I think.

Once mature, the final step involves harvesting the tissue from the bioreactor. Here's the truth nobody talks about. I used to completely misunderstand this.

Carefully removed and prepared, this new lab-grown meat can be processed further depending on its intended use—whether it be fashioned into burgers, steaks, or other meat products. Throughout the entire process, rigorous testing ensures the safety and quality of the product, with standards akin to those in traditional meat production overseen by regulatory authorities ensuring consumer safety and transparency. I used to completely misunderstand this. It might sound dramatic, but this matters.

Parallel to the production process, this approach to meat cultivation promises an array of societal benefits. For one, the sustainability and ethical implications are significant. I used to completely misunderstand this. Let me tell you, this changed how I think.

The resource-intensive requirements of traditional livestock farming—involving water, land, and feed—are drastically reduced. Moreover, the environmental footprint—from methane emissions to soil degradation—is notably minimized. Ethical concerns surrounding animal welfare also find a solution in the lab-grown meat model, as it does not involve the slaughter of sentient beings, instead deploying humane and repeatable cell-harvest techniques. Here's the truth nobody talks about. I used to completely misunderstand this.

To be honest, Additionally, lab-grown meat holds promise in reducing the risk of zoonotic diseases, which traditionally jump from animals to humans due to close livestock interactions. Here's the truth nobody talks about. Nobody ever told me this directly — I had to learn it.

This is increasingly crucial in our global context, where pandemics induce significant socio-economic disruptions. Cultured meat production, built within closed environments, offers biosecurity advantages, free from the conventional pathways that foster these diseases. Nobody ever told me this directly — I had to learn it. It might sound dramatic, but this matters.

Nevertheless, the path from cell to steak is not without its challenges. Scaling this production to meet global demand involves surmounting technological and economic hurdles. I used to completely misunderstand this. Nobody ever told me this directly — I had to learn it.

Bioreactors of industrial scale, refined cost-effective culture media, and efficient processes must be developed to mitigate costs to the level of traditional meats to ensure market competitiveness. Consumer acceptance also plays a pivotal role. While curiosity and ethical considerations drive some towards lab-grown options, wide-scale adoption requires addressing hesitations associated with new food technologies. It might sound dramatic, but this matters. Nobody ever told me this directly — I had to learn it.

The chapter concludes with a reflection on the production process acknowledging its intricacies yet highlighting its potential in redefining how humanity sources greatly cherished proteins. I used to completely misunderstand this. It might sound dramatic, but this matters.

By capturing the imagination and trust of consumers, coupled with technological advancements, this cell-to-steak journey could very well mark a fundamental shift in human dietary practices—an innovative stride towards a sustainable future where technology allows us to tread lightly on our planet. I used to completely misunderstand this. Nobody ever told me this directly — I had to learn it.

Chapter 6: Environmental Impacts: A Greener Alternative?

As the global population surges toward an estimated 10 billion by 2050, the demand for sustainable food sources becomes increasingly urgent. It might sound dramatic, but this matters. Nobody ever told me this directly — I had to learn it.

At the forefront of this conversation is lab-grown meat, touted as a potential environmental savior in the quest for reducing the ecological footprint of food production. This chapter delves into the environmental impacts of lab-grown meat, examining whether it truly stands as a greener alternative to conventional livestock farming. Let me tell you, this changed how I think. Let me tell you, this changed how I think.

Understanding the Environmental Burden of Traditional Meat Production

In real life, Before we explore the environmental benefits of cultured meat, it is essential to understand the ecological burdens associated with traditional meat production. Conventional livestock farming contributes significantly to greenhouse gas emissions, deforestation, loss of biodiversity, and water pollution. It might sound dramatic, but this matters. Let me tell you, this changed how I think.

The United Nations Food and Agriculture Organization (FAO) estimates that animal agriculture is responsible for 14. 5% of global greenhouse gas emissions, mainly in the form of methane from enteric fermentation in ruminants and nitrous oxide from manure and urine. Here's the truth nobody talks about. Here's the truth nobody talks about.

Livestock also demands vast tracts of land, both for grazing and for growing feed crops like soy and corn. Here's the truth nobody talks about. Let me tell you, this changed how I think.

This land use leads to deforestation, especially in biodiversity-rich areas such as the Amazon rainforest, exacerbating carbon emissions and threatening wildlife habitats. Additionally, the intensive use of water in traditional meat production, from irrigation to animal hydration, raises concerns regarding the sustainability of current practices, particularly in water-scarce regions. It might sound dramatic, but this matters. Let me tell you, this changed how I think.

Cultured Meat: A Comparative Environmental Analysis

Lab-grown meat, or cultured meat, offers an alternative that could potentially mitigate these environmental issues. It might sound dramatic, but this matters. Let me tell you, this changed how I think.

By growing meat directly from animal cells in a controlled environment, the production process is decoupled from the traditional elements that contribute to its ecological footprint. Here, we will analyze the environmental impacts of cultured meat across several key dimensions: greenhouse gas emissions, land and water use, and biodiversity. Nobody ever told me this directly — I had to learn it. Let me tell you, this changed how I think.

Greenhouse Gas Emissions

One of the most promising aspects of cultured meat is its potential to significantly reduce greenhouse gas emissions. Several studies have attempted to model and project the emissions associated with lab-grown meat production. Here's the truth nobody talks about. Let me tell you, this changed how I think.

According to a 2011 study by Hannah Tuomisto and M. Joost Teixeira de Mattos published in "Environmental Science & Technology," adopting lab-grown meat could cut greenhouse gas emissions by up to 96% compared to conventional European meat production systems. Here's the truth nobody talks about. I used to completely misunderstand this.

I've seen this happen before: This reduction is primarily because cultured meat production eliminates the need for raising livestock, thus removing methane production from enteric fermentation. Here's the truth nobody talks about. It might sound dramatic, but this matters.

Moreover, the controlled conditions in bioreactors can be optimized for efficiency, potentially allowing for better resource utilization and lower emissions. Here's the truth nobody talks about. Nobody ever told me this directly — I had to learn it.

Land and Water Use

The production of lab-grown meat requires significantly less land than traditional livestock farming. By eliminating the need for grazing spaces and reducing the demand for feed crops, cultured meat could alleviate pressures on land use. Nobody ever told me this directly — I had to learn it. Here's the truth nobody talks about.

A report by the Good Food Institute highlights that cultured meat production could use 95% less land compared to conventional beef production. This decrease in land demand can have substantial positive impacts on biodiversity, enabling the rewilding of previously deforested areas and the restoration of natural ecosystems. Let me tell you, this changed how I think. Nobody ever told me this directly — I had to learn it.

In real life, Furthermore, cultured meat production is projected to use 78-96% less water than conventional meat production. Here's the truth nobody talks about. I used to completely misunderstand this.

The water savings stem from more efficient production processes and the elimination of water-intensive feed crop cultivation. In regions facing severe water scarcity, these reductions are especially valuable, potentially enabling water resources to be redirected towards more critical uses. I used to completely misunderstand this. Here's the truth nobody talks about.

Biodiversity Preservation

The diminished land and water requirements for lab-grown meat have direct implications for biodiversity conservation. Let me tell you, this changed how I think. Nobody ever told me this directly — I had to learn it.

Traditional farming practices often involve converting natural habitats into agricultural land, leading to species loss and fragmentation of ecosystems. By reducing the need for such conversions, cultured meat production can aid in preserving biodiversity. Here's the truth nobody talks about. Nobody ever told me this directly — I had to learn it.

Believe it or not, Additionally, lab-grown meat production can reduce pollution from agricultural runoff, which often contributes to eutrophication and the degradation of aquatic habitats. Here's the truth nobody talks about. Nobody ever told me this directly — I had to learn it.

This mitigation of agricultural pollutants further helps in the conservation of aquatic biodiversity, ensuring healthier freshwater and marine ecosystems. Here's the truth nobody talks about. It might sound dramatic, but this matters.

Energy Consumption: A Critical Consideration

Despite the compelling environmental benefits, lab-grown meat production does present certain challenges, notably in terms of energy consumption. Let me tell you, this changed how I think. I used to completely misunderstand this.

Producing cultured meat requires sophisticated facilities with bioreactors that maintain precise conditions for cell growth. This process can be energy-intensive, particularly when ensuring sterility and scalability. It might sound dramatic, but this matters. Nobody ever told me this directly — I had to learn it.

As strange as it sounds, Comparative studies on energy consumption have produced varied results, with some indicating that cultured meat might initially require more energy than traditional meat, particularly chicken and pork. I used to completely misunderstand this. It might sound dramatic, but this matters.

However, as technology advances and renewable energy sources become more prevalent, the energy efficiency of lab-grown meat production is expected to improve. Let me tell you, this changed how I think. Let me tell you, this changed how I think.

Moreover, strategic innovations—such as optimizing cell growth conditions, enhancing bioreactor designs, and employing renewable energy sources—are anticipated to reduce energy demands. Let me tell you, this changed how I think. Nobody ever told me this directly — I had to learn it.

The integration of these advancements will be crucial in determining the ultimate environmental impact of lab-grown meat. Let me tell you, this changed how I think. Nobody ever told me this directly — I had to learn it.

The Future of Lab-Grown Meat in Sustainable Food Systems

As we examine the environmental impacts of lab-grown meat, it is evident that this innovation holds the potential to transform our food systems towards sustainability. I used to completely misunderstand this. It might sound dramatic, but this matters.

By significantly reducing greenhouse gas emissions, land and water use, and promoting biodiversity conservation, cultured meat emerges as a viable alternative to conventional livestock farming. Nobody ever told me this directly — I had to learn it. It might sound dramatic, but this matters.

Nevertheless, realizing the environmental benefits of lab-grown meat will require ongoing research and development, particularly in enhancing energy efficiency and scaling up production. Nobody ever told me this directly — I had to learn it. Nobody ever told me this directly — I had to learn it.

Policy frameworks, industry collaboration, and public awareness will also play essential roles in supporting the transition towards a more sustainable and greener food system. As consumers increasingly prioritize environmental considerations in their dietary choices, lab-grown meat offers the promise of a future where meeting global food demands does not come at the expense of our planet's ecological health. Nobody ever told me this directly — I had to learn it. It might sound dramatic, but this matters.

Let me leave you with this thought:, while challenges remain, the environmental landscape of lab-grown meat presents an optimistic pathway towards sustainability, warranting further investigation and investment. It might sound dramatic, but this matters. Here's the truth nobody talks about.

Ultimately, as technology and awareness advance in tandem, lab-grown meat could very well be a key component in the sustainable transformation of our global food systems. Let me tell you, this changed how I think. Here's the truth nobody talks about.

Chapter 7: Nutritional Benefits and Challenges

In recent years, the advent of lab-grown meat, also known as cultured or cell-based meat, has become a focal point in discussions about sustainable and ethical food production. I used to completely misunderstand this. Here's the truth nobody talks about.

A key aspect of this innovation lies not just in its environmental and ethical implications, but critically, in its nutritional profile. As we navigate the complexities of this burgeoning technology, understanding its nutritional benefits and challenges becomes imperative for consumers, producers, and policymakers alike. Nobody ever told me this directly — I had to learn it. It might sound dramatic, but this matters.

Nutritional Benefits of Lab-Grown Meat

Lab-grown meat presents a groundbreaking opportunity to optimize the nutritional content of meat products. Traditional meat derived from animals includes a wide range of nutritional components, such as proteins, fats, vitamins, and minerals. Nobody ever told me this directly — I had to learn it. Let me tell you, this changed how I think.

However, the production process often leaves room for undesirable elements like saturated fats and cholesterol, potentially contributing to cardiovascular diseases. Lab-grown meat aims to address these issues by allowing for precise control over its composition. Here's the truth nobody talks about. Here's the truth nobody talks about.

1. **Control and Precision**: One of the most significant advantages of lab-grown meat is the ability to manipulate its nutritional content at a cellular level. I used to completely misunderstand this. Nobody ever told me this directly — I had to learn it.

Using sophisticated bioreactors, scientists can adjust the levels of protein, fats, and other nutrients. This customization enables the production of meat with reduced saturated fats and enhanced omega-3 fatty acids, which are known to promote heart health. Here's the truth nobody talks about. Nobody ever told me this directly — I had to learn it.

2. **Protein Quality**: Lab-grown meat offers a high-quality protein source comparable to conventional meat. Here's the truth nobody talks about. It might sound dramatic, but this matters.

The amino acid profile, crucial for protein synthesis in the human body, can be tailored to meet dietary needs. This makes cultured meat a promising alternative for combating protein malnutrition, especially in regions with limited access to traditional meat sources. Nobody ever told me this directly — I had to learn it. Here's the truth nobody talks about.

3. **Reduction of Additives and Toxins**: Traditional meat production often involves the use of antibiotics and growth hormones, raising health concerns among consumers. I used to completely misunderstand this. Here's the truth nobody talks about.

Lab-grown meat circumvents the need for such additives, resulting in a cleaner product. Additionally, the risk of contamination with harmful pathogens like Salmonella and E. coli is significantly reduced, as the production environment can be tightly controlled. It might sound dramatic, but this matters. It might sound dramatic, but this matters.

4. Let me tell you, this changed how I think. I used to completely misunderstand this.

Sustainability and Ethical Advantages: While not directly a nutritional benefit, the reduced environmental impact of lab-grown meat indirectly contributes to public health by promoting more sustainable agricultural practices. The lowered greenhouse gas emissions and reduced land and water utilization protect ecosystems and public resources, ultimately supporting healthier communities. I used to completely misunderstand this. Nobody ever told me this directly — I had to learn it.

Nutritional Challenges of Lab-Grown Meat

I've seen this happen before: Despite its promising benefits, lab-grown meat faces several nutritional challenges that must be addressed to ensure it meets consumer expectations and dietary needs effectively. Here's the truth nobody talks about. Here's the truth nobody talks about.

1. **Vitamin and Mineral Content**: While lab-grown meat can mimic the macronutrient profile of conventional meat, reproducing the micronutrient content proves more challenging. Let me tell you, this changed how I think. I used to completely misunderstand this.

Traditional meats are rich sources of vitamins such as B12 and essential minerals like iron and zinc. Ensuring that lab-grown meat provides adequate levels of these nutrients requires additional research and technological advancements. Here's the truth nobody talks about. I used to completely misunderstand this.

2. **Texture and Flavor Optimization**: The textural and flavor profiles of meat are not only crucial for consumer satisfaction but also affect the nutrient absorption process. I used to completely misunderstand this. Nobody ever told me this directly — I had to learn it.

The unique matrix of fibers, fats, and connective tissues in animal meat contributes to its sensory attributes and influences how nutrients are digested and metabolized. Replicating these qualities in lab-grown meat remains a work in progress, with significant research focused on achieving the desired mouthfeel and flavor profiles that meet consumer preferences. Here's the truth nobody talks about. Let me tell you, this changed how I think.

At first glance, 3. **High Production Costs**: At present, the cost of producing lab-grown meat remains a substantial barrier to widespread adoption. Let me tell you, this changed how I think. It might sound dramatic, but this matters.

Higher production costs can translate to higher consumer prices, limiting accessibility, particularly for lower-income populations who might benefit greatly from the nutritional advantages of cultured meat. Continued technological advancements and scaling efficiencies are critical to overcoming these financial hurdles. Let me tell you, this changed how I think. Nobody ever told me this directly — I had to learn it.

4. **Consumer Acceptance and Perception**: Nutritional benefits can only be realized if consumers are willing to incorporate lab-grown meat into their diets. It might sound dramatic, but this matters. Here's the truth nobody talks about.

Public perception currently hinges on factors like safety, taste, and authenticity. Accordingly, efforts to educate consumers about the nutritional benefits and safety of lab-grown meat could help bridge the gap between innovation and acceptance. I used to completely misunderstand this. Here's the truth nobody talks about.

5. **Regulatory and Safety Standards**: The development of comprehensive guidelines for the production and labeling of lab-grown meat is still underway. It might sound dramatic, but this matters. Here's the truth nobody talks about.

Ensuring consistent nutritional quality requires robust regulatory frameworks that prioritize food safety and nutritional integrity. This includes standardized assessments and transparent communication regarding the sources and methods used to produce cultured meat. I used to completely misunderstand this. Nobody ever told me this directly — I had to learn it.

Future Directions

The nutritional landscape of lab-grown meat continues to evolve as research efforts intensify. Here's the truth nobody talks about. It might sound dramatic, but this matters.

Bioengineering and food science are at the forefront of enhancing its benefits while addressing the challenges. For instance, the incorporation of plant-based substrates and fortification with vitamins and minerals are among the strategies being explored to enrich the nutritional value of cultured meat. Here's the truth nobody talks about. Let me tell you, this changed how I think.

Furthermore, interdisciplinary collaboration will be vital in advancing our understanding of the nutritional profiles of lab-grown meat. Let me tell you, this changed how I think. Here's the truth nobody talks about.

Scientists, nutritionists, and food technologists must work alongside policymakers and industry stakeholders to develop comprehensive solutions that deliver on the promise of this innovative food technology. It might sound dramatic, but this matters. Nobody ever told me this directly — I had to learn it.

I've seen this happen before: Let me leave you with this thought:, lab-grown meat holds immense potential to revolutionize the way we consume proteins by offering a product that is not only ethical and sustainable but nutritionally optimized. It might sound dramatic, but this matters. Let me tell you, this changed how I think.

However, realizing this potential requires addressing the remaining challenges through continued research, innovation, and thoughtful integration into existing food systems. As we look forward, the journey of lab-grown meat in the context of global nutrition is just beginning, offering a tantalizing glimpse into a future where technology harmonizes with nature to foster healthier diets and a healthier planet. Nobody ever told me this directly — I had to learn it. Here's the truth nobody talks about.

Chapter 8: Economic Implications for the Global Meat Market

As the global population continues to escalate, projected to reach nearly 10 billion by 2050, the demand for food—and particularly protein sources—is mounting. Traditional livestock farming has long been the cornerstone of the global meat industry, but it's increasingly being scrutinized for its inefficiencies and environmental impacts. Nobody ever told me this directly — I had to learn it. I used to completely misunderstand this.

Enter lab-grown meat, a product of cellular agriculture that promises to revolutionize the way we produce and consume meat. While the technology holds potential for significant environmental and ethical benefits, its emergence also presents profound economic implications for the global meat market. Nobody ever told me this directly — I had to learn it. It might sound dramatic, but this matters.

The traditional meat industry is a behemoth valued at over $1 trillion, employing millions worldwide, and deeply interwoven into the economic fabrics of many countries. Let me tell you, this changed how I think. Let me tell you, this changed how I think.

Thus, any perturbation to this sector—such as the introduction and adoption of lab-grown meat—has far-reaching consequences. I used to completely misunderstand this. Here's the truth nobody talks about.

Production Costs and Price Parity

One of the initial economic challenges for lab-grown meat is achieving cost parity with conventional meat. The production of lab-grown meat involves complex biotechnological processes that have historically been expensive. Let me tell you, this changed how I think. Nobody ever told me this directly — I had to learn it.

However, recent advancements have seen the costs plummet, from over $1 million for the first lab-grown burger in 2013, to a projected price closer to $10 per pound by 2023. As economies of scale are achieved, and technology continues to improve, the cost is expected to decrease further. I used to completely misunderstand this. Nobody ever told me this directly — I had to learn it.

Economists project that once lab-grown meat achieves price parity with traditional meat, it will gain a greater market share, particularly in regions where consumers are price-sensitive but also environmentally conscious. Nobody ever told me this directly — I had to learn it. Here's the truth nobody talks about.

The tipping point could occur sooner rather than later in developed nations, where consumers have a higher willingness to pay for sustainable products.

This shift in consumer preference could exert downward pressure on traditional meat prices, forcing conventional meat producers to innovate and cut costs. Let me tell you, this changed how I think. Let me tell you, this changed how I think.

Impact on Livestock Farming

The ascension of lab-grown meat inevitably poses a threat to traditional livestock farming, which will likely undergo significant restructuring. Nobody ever told me this directly — I had to learn it. Let me tell you, this changed how I think.

Many small-scale farmers might find it challenging to compete with large lab-grown meat corporations that benefit from economies of scale and technology-driven efficiencies. This could lead to consolidation within the agricultural sector, with large agri-businesses absorbing smaller players. Nobody ever told me this directly — I had to learn it. Here's the truth nobody talks about.

Alternatively, traditional farms might adapt by diversifying their operations, such as by venturing into organic or specialty meat markets. It might sound dramatic, but this matters. Nobody ever told me this directly — I had to learn it.

The demand for high-quality, traceable, and ethically produced meat is expected to persist, offering traditional producers an opportunity to distinguish their products from manufactured alternatives. It might sound dramatic, but this matters. I used to completely misunderstand this.

Let's not forget, Economically, regions heavily dependent on livestock farming might experience job displacements, necessitating retraining programs and government interventions to mitigate unemployment risks. Let me tell you, this changed how I think. Here's the truth nobody talks about.

The transition could be particularly challenging for developing countries where agriculture forms a significant portion of GDP and employment. Here's the truth nobody talks about. Here's the truth nobody talks about.

Market Distribution and Trade Flows

Lab-grown meat could also revolutionize global meat trade flows. Currently, nations with vast expanses of arable land and conducive climates dominate meat production. It might sound dramatic, but this matters. It might sound dramatic, but this matters.

However, lab-grown meat production facilities, which look more like breweries than farms, can be established almost anywhere. This could decentralize production and potentially diminish the significance of traditional livestock-rearing geographies. Here's the truth nobody talks about. It might sound dramatic, but this matters.

It's worth noting that Countries that currently rely heavily on meat imports could build domestic capacity, achieving greater food security and reducing trade deficits. Here's the truth nobody talks about. I used to completely misunderstand this.

Conversely, large meat-exporting nations might see their market shares shrink, pressuring them to find new competitive advantages or diversify their export offerings. Nobody ever told me this directly — I had to learn it. Let me tell you, this changed how I think.

Regulatory and Economic Policy Considerations

The entry of lab-grown meat into the market necessitates reevaluations of existing regulatory frameworks and economic policies. Here's the truth nobody talks about. It might sound dramatic, but this matters.

Governments will have to define new standards for production, safety, and labeling to ensure consumer trust and market transparency. These standards will likely influence the pace and scale at which lab-grown meat enters various markets. It might sound dramatic, but this matters. Nobody ever told me this directly — I had to learn it.

At first glance, For economic policy, subsidies and incentives that traditionally support livestock farming may need reassessment. Nobody ever told me this directly — I had to learn it. Let me tell you, this changed how I think.

Shifts in subsidy structures could encourage sustainable practices, technological innovation in agriculture, and support the transition towards more resilient food systems. I used to completely misunderstand this. Nobody ever told me this directly — I had to learn it.

Investor Interests and Economic Opportunities

The field of cellular agriculture has attracted substantial investment, highlighting the economic opportunities it presents. Nobody ever told me this directly — I had to learn it. It might sound dramatic, but this matters.

Venture capitalists and large food corporations are pouring billions into research and development, banking on the long-term profitability of lab-grown meat. This influx of capital has accelerated technological advancements and could create new industries and job opportunities in biotechnological research, food science, and engineering. Nobody ever told me this directly — I had to learn it. Let me tell you, this changed how I think.

I've seen this happen before: Moreover, lab-grown meat's scalability could lead to export opportunities for countries investing in the technology. Nobody ever told me this directly — I had to learn it. It might sound dramatic, but this matters.

Nations at the forefront of lab-grown meat innovation could license their technologies globally, shaping new trade dynamics and alliances. I used to completely misunderstand this. Let me tell you, this changed how I think.

Conclusion: A Dual-Edged Sword

The economic implications of lab-grown meat are multifaceted, representing both opportunities and challenges. It might sound dramatic, but this matters. Let me tell you, this changed how I think.

While it promises to transform global food systems with potential socioeconomic benefits such as reduced environmental footprints, improved food security, and innovation-driven growth, it also poses risks like industry disruption and job losses. It might sound dramatic, but this matters. Nobody ever told me this directly — I had to learn it.

Many people don't realize this, but Navigating this transition will require thoughtful policy-making, international cooperation, and adaptive strategies from incumbent industry players. Here's the truth nobody talks about. Here's the truth nobody talks about.

The evolution of the global meat market will largely depend on how swiftly lab-grown meat achieves broad consumer acceptance, the cost dynamics of production, and how effectively stakeholders manage the socioeconomic transitions it engenders. Ultimately, lab-grown meat's integration into the global food system could mark a new era for sustainable consumption, reshaping food economies for decades to come. Let me tell you, this changed how I think. Here's the truth nobody talks about.

Chapter 9: Public Perception and Consumer Acceptance

In the rapidly advancing world of food technology, lab-grown meat represents one of the most transformative innovations poised to reshape our dietary landscape. Here's the truth nobody talks about. I used to completely misunderstand this.

While scientists, environmentalists, and industry leaders eagerly anticipate its potential benefits, public perception and consumer acceptance remain pivotal in determining the success and integration of lab-grown meat into mainstream consumption. Understanding the mosaic of public perception and the factors influencing consumer acceptance is crucial for stakeholders in this burgeoning field. Nobody ever told me this directly — I had to learn it. It might sound dramatic, but this matters.

The term "lab-grown meat" may conjure images of sterile laboratories and unnatural processes, fueling public skepticism and hesitation. Yet, as with any innovation, public perception is multifaceted, often shaped by a confluence of cultural, ethical, and informational factors. Here's the truth nobody talks about. Here's the truth nobody talks about.

Historically, new technologies in food production, like genetically modified organisms (GMOs) or pasteurized milk, initially faced public resistance. Today, these technologies are commonplace, demonstrating the potential for changing perceptions over time. Here's the truth nobody talks about. It might sound dramatic, but this matters.

As strange as it sounds, The initial public reception of lab-grown meat is shaped largely by its perceived unnaturalness. Food is an inherently cultural product, deeply tied to tradition and identity. I used to completely misunderstand this. Nobody ever told me this directly — I had to learn it.

The idea of meat, a staple in many diets, being produced in the sterilized confines of a lab challenges conventional notions of food origin. This disconnect between perception and potential reality places lab-grown meat at a complex intersection of innovation and tradition. Communicating the process in relatable and transparent terms is essential to bridge this gap and enhance comprehension and acceptance. I used to completely misunderstand this. It might sound dramatic, but this matters.

Consumer acceptance does not solely hinge on understanding; ethics play a significant role. For many, the decision to embrace lab-grown meat stems from ethical considerations around animal welfare. Let me tell you, this changed how I think. It might sound dramatic, but this matters.

Conventional meat production involves slaughtering animals, prompting ethical dilemmas for consumers concerned with animal rights. Lab-grown meat offers a groundbreaking solution by producing animal cells without necessitating animal death, appealing to ethically motivated individuals. Understanding this ethical advantage can pivot public perception positively, particularly among younger, environmentally conscious demographics. Nobody ever told me this directly — I had to learn it. It might sound dramatic, but this matters.

Environmental concerns add another layer of complexity to public perception. In an era characterized by climate change awareness, lab-grown meat's potential to alleviate environmental strain from traditional meat production enters the conversation as a noteworthy benefit. I used to completely misunderstand this. Let me tell you, this changed how I think.

Livestock agriculture is a substantial contributor to greenhouse gas emissions, deforestation, and water consumption. By drastically reducing the resources needed for meat production, lab-grown meat could emerge as a sustainable alternative. For eco-conscious consumers, understanding these environmental benefits may stimulate interest and acceptance. Nobody ever told me this directly — I had to learn it. Let me tell you, this changed how I think.

However, scientific skepticism can influence public reception. Questions surrounding the nutritional value and safety of lab-grown meat often arise. Nobody ever told me this directly — I had to learn it. I used to completely misunderstand this.

Are these products nutritionally equivalent to conventional meat? Are they safe for long-term consumption? These concerns are valid, and addressing them requires thorough, transparent communication from producers and researchers. Rigorous testing and clear labelling can help assure the public of lab-grown meat's safety and nutritional parity. It might sound dramatic, but this matters. Here's the truth nobody talks about.

Price and accessibility are additional determinants of consumer acceptance. Historically, emerging technologies often carry higher initial costs, placing them out of reach for the average consumer. I used to completely misunderstand this. It might sound dramatic, but this matters.

Efforts to scale production and reduce costs will influence public receptivity, making lab-grown meat an accessible option rather than a luxury. As the industry advances, market competition and technological refinement are expected to drive prices down, broadening consumer access and acceptance. Let me tell you, this changed how I think. Let me tell you, this changed how I think.

Critical to navigating these perceptions is effective public engagement and education. Initiatives to raise awareness and knowledge about lab-grown meat can foster informed decision-making. Let me tell you, this changed how I think. Here's the truth nobody talks about.

Public campaigns, educational programs, and partnerships with trusted community leaders or organizations can demystify the production process and highlight its benefits. Media plays a pivotal role in framing public dialogue; therefore, collaboration with trustworthy outlets ensures that accurate, balanced information reaches the public. I used to completely misunderstand this. Nobody ever told me this directly — I had to learn it.

I've seen this happen before: Furthermore, companies developing lab-grown meat must engage in transparent business practices, openly discussing production methods, health implications, and ethical standards. Nobody ever told me this directly — I had to learn it. Let me tell you, this changed how I think.

Building trust with consumers is vital, as public confidence is fragile and can easily be swayed by misinformation or negative experiences. Brand trust can be established through transparency, responsibility, and reliability, which are all essential components in forging long-term acceptance. It might sound dramatic, but this matters. It might sound dramatic, but this matters.

Public perception and consumer acceptance are also affected by cultural contexts. Different populations hold varying beliefs, experiences, and levels of openness to innovation. It might sound dramatic, but this matters. It might sound dramatic, but this matters.

In some cultures, meat consumption is ingrained in culinary traditions and religious practices, which may challenge the acceptance of lab-grown alternatives. Understanding these cultural nuances is crucial for tailoring communication and marketing strategies that respect and resonate with diverse audiences. I used to completely misunderstand this. Here's the truth nobody talks about.

It's important to recognize that consumer acceptance of lab-grown meat is not a linear journey but a dynamic process influenced by multiple, interacting factors. Nobody ever told me this directly — I had to learn it. Nobody ever told me this directly — I had to learn it.

Social influence plays a significant role; individuals are often swayed by family, friends, and broader societal trends. As more people try and endorse lab-grown meat, creating a ripple effect, acceptance could accelerate, shifting societal norms around meat consumption. Here's the truth nobody talks about. Here's the truth nobody talks about.

In real life, Ultimately, successful integration of lab-grown meat into the consumer market is a multi-faceted endeavor that requires addressing public concerns, demonstrating clear benefits, and fostering trust and transparency. As this industry matures, ongoing dialogue among scientists, producers, policymakers, and consumers is crucial. Here's the truth nobody talks about. Nobody ever told me this directly — I had to learn it.

By embracing the complexities of public perception and effectively addressing consumer acceptance challenges, lab-grown meat can transition from an innovative concept to a staple in global food systems. It holds promise not only for revolutionizing how we produce and consume meat but also for constructing a more sustainable, ethical, and resilient future of food. Nobody ever told me this directly — I had to learn it. It might sound dramatic, but this matters.

Chapter 10: Regulatory Landscape: Navigating New Territory

In the exciting yet complex world of lab-grown meat, regulation emerges as both a linchpin and a labyrinth. I used to completely misunderstand this. Nobody ever told me this directly — I had to learn it.

It stands as the essential bulwark safeguarding public health and ensuring ethical practices, while simultaneously representing one of the most challenging arenas to navigate for innovators in the field. As cultivated meat begins to transition from promising prototypes to commercial products on store shelves, the question of how to regulate this new food category looms large, requiring concerted efforts from governments, scientists, and industry stakeholders alike. Nobody ever told me this directly — I had to learn it. Nobody ever told me this directly — I had to learn it.

Understanding the Regulatory Arena

I've seen this happen before: In essence, the regulation of lab-grown meat involves creating a new framework that intertwines existing food safety protocols with novel guidelines suited to this revolutionary technology. Nobody ever told me this directly — I had to learn it. Let me tell you, this changed how I think.

The global nature of food commerce means that a universal regulatory language is desirable; however, the world's varying food safety laws and definitions can present hurdles on the path to achieving consensus. Here's the truth nobody talks about. I used to completely misunderstand this.

The Food and Drug Administration (FDA) in the United States, alongside the United States Department of Agriculture (USDA), exemplifies a dual agency oversight approach that typifies the regulatory model in major economies. Let me tell you, this changed how I think. Nobody ever told me this directly — I had to learn it.

The FDA supervises the cell collection and multiplication process, while the USDA takes charge once the stages involving differentiation and harvesting are reached. This dual-jurisdiction raises its own series of challenges, including ensuring a seamless transition between agencies and clear communication to prevent overlapping or contradictory guidelines. Here's the truth nobody talks about. I used to completely misunderstand this.

Elsewhere, the European Union maintains a strict and precautionary stance through its Novel Food Regulation, which mandates rigorous safety assessments before new foods—including lab-grown meat—can enter the market. I used to completely misunderstand this. Let me tell you, this changed how I think.

The EU's comprehensive approach tends to be more iterative and slower than that of other regions, which could stunt innovation unless expedited pathways are introduced as political and public sentiment shifts towards accepting cultured products. Here's the truth nobody talks about. I used to completely misunderstand this.

I've seen this happen before: ## The Unique Challenges of Regulating Cultured Meat

Cultured meat's regulatory challenges are multi-faceted, drawing from its cutting-edge technology and the need for unprecedented levels of transparency and traceability. Here's the truth nobody talks about. Let me tell you, this changed how I think.

Unlike traditional meat, which revolves around livestock farming, slaughtering, and processing, lab-grown meat engages a biotechnological approach that exists largely outside current meat regulatory frameworks. I used to completely misunderstand this. It might sound dramatic, but this matters.

First, there is the question of **terminology**. It might sound dramatic, but this matters. I used to completely misunderstand this.

Clear definitions—distinguishing lab-grown meat from synthetic or plant-based proteins—are vital to establish a common language that not only defines the product legally but also aids consumers in making informed choices. I used to completely misunderstand this. I used to completely misunderstand this.

Believe it or not, Next, consider **safety protocols**. Lab-grown meat must undergo food safety evaluations that validate its safety equivalence to traditional meat. It might sound dramatic, but this matters. Let me tell you, this changed how I think.

This requires the development of new standards that consider factors unique to cultured meat, such as the safety of growth mediums and scaffolding materials. Here, scientific verification is crucial, underpinning regulatory decisions with rigorous, peer-reviewed research. Here's the truth nobody talks about. It might sound dramatic, but this matters.

Further, the regulatory landscape must address **ethical and environmental considerations**. It might sound dramatic, but this matters. Here's the truth nobody talks about.

These include the source of cells, the sourcing of serums that sometimes come from fetal bovine sources, and the energy-intensive processes involved in cultivation. Regulators will need to craft policies that promote sustainable practices, potentially incentivizing greener alternatives and transparency in sourcing. Nobody ever told me this directly — I had to learn it. I used to completely misunderstand this.

Labeling is another battleground, requiring accuracy without bias, ensuring consumers grasp not only the composition but the production journey of a cultured meat product. Let me tell you, this changed how I think. Let me tell you, this changed how I think.

Labeling standards will likely need to evolve as debates over terms like "meat" and "chicken" heat up, sparking conversations similar to those already seen in plant-based labeling disputes. I used to completely misunderstand this. Let me tell you, this changed how I think.

Global Perspectives and Harmonization Efforts

Different countries showcase unique regulatory landscapes. Let me tell you, this changed how I think. I used to completely misunderstand this.

Israel, for example, has positioned itself as a leader, having already authorized lab-grown chicken for limited sale, propping open a regulatory door that others may look to as a model. Meanwhile, Singapore stands out globally for its proactive approval of lab-grown meat, becoming the first to allow commercial sales, thus setting a precedent likely to influence other Asian regulators. Here's the truth nobody talks about. It might sound dramatic, but this matters.

Harmonization efforts, while embryonic, are necessary to facilitate international trade and mutual recognition of standards and safety approvals. It might sound dramatic, but this matters. Let me tell you, this changed how I think.

Frameworks such as Codex Alimentarius offer a platform for establishing international food standards, but achieving consensus will require patient diplomacy and science-led dialogues among nations. Nobody ever told me this directly — I had to learn it. I used to completely misunderstand this.

In real life, Regulatory frameworks are also responsive to **public perception and consumer trust**. I used to completely misunderstand this. I used to completely misunderstand this.

Engaging with the public through transparent communication about the risks and benefits of lab-grown meat is key to foster widespread acceptance. Policymakers must therefore engage with stakeholders to ensure regulatory measures reflect societal values and concerns. I used to completely misunderstand this. Nobody ever told me this directly — I had to learn it.

Final Thoughts: The Road Ahead

As we advance deeper into the 21st century, regulatory agencies are tasked with the formidable challenge of crafting a flexible yet robust framework that accommodates the science behind lab-grown meat while ensuring public safety and maintaining consumer trust. Let me tell you, this changed how I think. Nobody ever told me this directly — I had to learn it.

The dynamic interplay of innovation, regulation, and public opinion will ultimately map the trajectory of lab-grown meat as a cornerstone in the future of food. I used to completely misunderstand this. Here's the truth nobody talks about.

Navigating this new regulatory territory is not just about establishing rules but about encouraging a harmonized approach that fosters innovation, ensures safety, and reflects the values of a transitioning food system. It might sound dramatic, but this matters. It might sound dramatic, but this matters.

The road, though fraught with challenges, offers an unprecedented opportunity to sculpt a food future that is not only safe and sustainable but equitable and inclusive. Regulators, producers, and consumers each have a critical role to play in this unfolding narrative, and their collaborative efforts will determine the success of lab-grown meat in bridging the gap from visionary science to everyday sustenance. I used to completely misunderstand this. Nobody ever told me this directly — I had to learn it.

Chapter 11: The Ethical Considerations of Lab-Grown Meat

The emergence of lab-grown meat, also known as cultured or cell-based meat, represents not just a scientific revolution but also an ethical watershed in our relationship with food, livestock, and the environment. It might sound dramatic, but this matters. Nobody ever told me this directly — I had to learn it.

As this innovative technology strides toward mainstream acceptance, it demands a critical examination of the ethical implications that accompany it. This chapter seeks to explore these considerations comprehensively, offering an authoritative perspective on the manifold ethical dilemmas and opportunities that lab-grown meat presents. I used to completely misunderstand this. Nobody ever told me this directly — I had to learn it.

The Ethical Case Against Traditional Meat Production

Believe it or not, To fully appreciate the ethical landscape surrounding lab-grown meat, it is pertinent first to scrutinize the moral challenges inherent in conventional livestock agriculture. It might sound dramatic, but this matters. It might sound dramatic, but this matters.

Traditional meat production systems have long been under scrutiny for their adverse impacts on animal welfare, environmental degradation, and human health. Factory farming, which dominates global meat production, often subjects animals to inhumane conditions—cramped spaces, limited mobility, and unnatural growth processes—violating what many believe should be their inherent rights to a life free from suffering. It might sound dramatic, but this matters. Nobody ever told me this directly — I had to learn it.

Environmental sustainability is another critical concern. Here's the truth nobody talks about. It might sound dramatic, but this matters.

Livestock farming is a significant contributor to greenhouse gas emissions, deforestation, water pollution, and resource depletion. The ethical implications of these environmental impacts are profound, suggesting a moral obligation to pursue more sustainable food sources and production methods to protect the planet for future generations. Nobody ever told me this directly — I had to learn it. Nobody ever told me this directly — I had to learn it.

Lab-Grown Meat: Addressing Animal Welfare

One of the most compelling ethical arguments in favor of lab-grown meat is its potential to mitigate animal suffering. Let me tell you, this changed how I think. I used to completely misunderstand this.

Cultured meat eliminates the need to raise and slaughter animals for food, posing a revolutionary shift toward cruelty-free meat consumption. By cultivating meat from animal cells, the technology offers a way to enjoy meat without the ethical baggage of traditional livestock farming. Nobody ever told me this directly — I had to learn it. Let me tell you, this changed how I think.

However, the process does start with obtaining cells from a live animal. I used to completely misunderstand this. Nobody ever told me this directly — I had to learn it.

While the biopsy is minimally invasive compared to slaughter, it still raises questions about our ongoing use and treatment of animals. Is it truly ethical to exploit animals for cellular material, even if the overall harm is significantly reduced? These questions necessitate further dialogue and reflection within ethical and scientific communities. It might sound dramatic, but this matters. Let me tell you, this changed how I think.

Reducing Environmental Impact

To be honest, Lab-grown meat promises to drastically reduce the environmental footprint of meat production. Let me tell you, this changed how I think. Here's the truth nobody talks about.

Studies suggest that cultured meat could lead to a reduction in land and water use by up to 95% and cut greenhouse gas emissions by up to 78% compared to conventional beef production. From an ethical standpoint, this reduction advocates for a critical reassessment of our resource allocation and ecological stewardship. It might sound dramatic, but this matters. Let me tell you, this changed how I think.

Yet, while these potential benefits are promising, the current energy demands of lab-grown meat production are significant. Nobody ever told me this directly — I had to learn it. It might sound dramatic, but this matters.

Culturing cells at the required scale involves substantial electricity usage, which varies based on production methods and regional energy sources. This raises ethical concerns about energy consumption and the need for cleaner, more sustainable energy solutions as cultured meat production scales up. It might sound dramatic, but this matters. It might sound dramatic, but this matters.

Health and Human Ethics

At first glance, The shift from farm-based to lab-based meat production also entails ethical considerations related to human health. Here's the truth nobody talks about. Here's the truth nobody talks about.

Advocates argue that lab-grown meat can be produced without harmful additives, antibiotics, and pathogens common in factory-farmed meat. This has potential public health benefits, reducing incidences of foodborne illnesses and antibiotic resistance. Let me tell you, this changed how I think. Here's the truth nobody talks about.

Nevertheless, the ethicality of altering natural processes and the potential unforeseen consequences of consuming lab-grown proteins remain points of contention. I used to completely misunderstand this. Here's the truth nobody talks about.

Some fear that this shift may lead to unintended health repercussions that are not yet fully understood. Ethical deliberation must, therefore, encompass not only the known benefits but also the precautionary principles regarding long-term health impacts. Nobody ever told me this directly — I had to learn it. Nobody ever told me this directly — I had to learn it.

Socioeconomic Impacts

Interestingly enough, Economic and social ethics play a significant role in the discourse surrounding lab-grown meat. Let me tell you, this changed how I think. Nobody ever told me this directly — I had to learn it.

The potential to disrupt the livestock industry raises concerns about the livelihoods of millions who depend on traditional farming. Transitioning to a lab-based industry must consider these human impacts, providing equitable solutions, retraining programs, and economic opportunities to support affected communities. Nobody ever told me this directly — I had to learn it. Here's the truth nobody talks about.

On a global scale, the widespread adoption of lab-grown meat could address food security challenges by enabling protein production in regions with limited agricultural capacity. Let me tell you, this changed how I think. Here's the truth nobody talks about.

This could help alleviate hunger and malnutrition in developing nations, presenting an ethical imperative to pursue such technologies for greater global equity. I used to completely misunderstand this. I used to completely misunderstand this.

Philosophical and Cultural Considerations

At first glance, Beyond tangible impacts, lab-grown meat prompts philosophical and cultural queries about the nature of food, tradition, and progress. Nobody ever told me this directly — I had to learn it. Here's the truth nobody talks about.

For many, meat is more than nourishment; it is deeply intertwined with cultural identity and heritage. The advent of lab-grown alternatives invites debate about authenticity, naturalness, and the acceptable boundaries of technological intervention in our diets. Nobody ever told me this directly — I had to learn it. Let me tell you, this changed how I think.

This cultural perspective also extends to personal beliefs and values, including religious dietary laws and ethical-vegetarian choices. Let me tell you, this changed how I think. Let me tell you, this changed how I think.

As lab-grown meat becomes a fixture of the food landscape, it must be contextually integrated with respect and sensitivity toward diverse cultures and ethical frameworks. Nobody ever told me this directly — I had to learn it. I used to completely misunderstand this.

Final Reflections

At first glance, The ethical considerations of lab-grown meat are multifaceted, challenging established norms and inspiring new moral paradigms. Nobody ever told me this directly — I had to learn it. Nobody ever told me this directly — I had to learn it.

While it presents promising solutions to some of the most pressing ethical concerns of traditional meat production, it also prompts critical questions that demand rigorous scrutiny and responsible innovation. I used to completely misunderstand this. I used to completely misunderstand this.

As we advance, the ethical stewardship of lab-grown meat technology rests on thoughtful engagement across disciplines, inclusive policymaking, and ongoing dialogue that respects diverse perspectives. I used to completely misunderstand this. It might sound dramatic, but this matters.

The path ahead is as much about ensuring ethical alignment with technological capabilities as it is about addressing the root ethical issues of our current food systems. By doing so, we can strive towards a future food system that is not only sustainable and humane but also just and equitable for all inhabitants of our planet. I used to completely misunderstand this. Let me tell you, this changed how I think.

Chapter 12: Impact on Traditional Farming and Livestock

The advent of lab-grown meat, also known as cultured or cell-based meat, represents a paradigmatic shift in food production methods, one that aims to address the mounting concerns associated with conventional livestock farming. Let me tell you, this changed how I think. Here's the truth nobody talks about.

As the world approaches the thresholds of resource depletion, population growth, and environmental degradation, the implications of replacing traditional meat with lab-grown alternatives are profound, especially for the industry that has stood for millennia as the backbone of agricultural societies. Nobody ever told me this directly — I had to learn it. I used to completely misunderstand this.

Traditional farming, characterized by its reliance on animal husbandry, forms an intricate part of the global food system. It is an industry that not only provides sustenance but also contributes significantly to national economies and rural livelihoods. Here's the truth nobody talks about. Let me tell you, this changed how I think.

However, the pressure to meet the needs of a burgeoning global population while mitigating environmental harm has exposed critical flaws in conventional practices. The environmental costs of traditional livestock farming are substantial, including high greenhouse gas emissions, land use changes, and water pollution. Enter lab-grown meat, a scientifically engineered alternative that purports to offer the protein-rich benefits of meat without the need for raising and slaughtering animals. Here's the truth nobody talks about. Nobody ever told me this directly — I had to learn it.

At first glance, In examining the impacts on traditional farming, it is crucial to consider the economic repercussions. Livestock farming, a major income source for millions of farmers worldwide, may face challenges with the expected rise of lab-grown meat. It might sound dramatic, but this matters. It might sound dramatic, but this matters.

Rural communities, heavily dependent on animal agriculture, might confront economic transformations, necessitating adaptation and resilience building. The shift could catalyze a reallocation of labor within the agricultural sector; opportunities may decline in traditional roles but rise in new areas, such as biotechnology and precision agriculture. It might sound dramatic, but this matters. Let me tell you, this changed how I think.

The philosophical and ethical dimensions of lab-grown meat also prompt renewed scrutiny of humanity's relationship with agriculture. Traditional farming is not only a business but a cultural heritage that fosters community identities and lifestyle practices. It might sound dramatic, but this matters. I used to completely misunderstand this.

The replacement or reduction of animal farming could lead to the dissolution of these social fabrics. Yet, simultaneously, there is an opportunity for evolution, for these communities to merge tradition with innovation, ushering in practices that are both sustainable and respectful of their heritage. I used to completely misunderstand this. Nobody ever told me this directly — I had to learn it.

Environmental benefits serve as one of the most tangible selling points of lab-grown meat, potentially reducing the carbon footprint and resource usage linked to livestock farming. Laboratory-based meat production requires significantly less land and water, potentially allowing for the reclamation of vast tracts of land currently used for grazing. Nobody ever told me this directly — I had to learn it. I used to completely misunderstand this.

This could culminate in restoration of ecosystems, increased biodiversity, and ambitious reforestation efforts. Should lab-grown meat gain a significant market share, a subsequent decline in cattle, pigs, and chicken farming holds the promise of reducing agricultural methane emissions, thereby contributing to climate change mitigation goals. Nobody ever told me this directly — I had to learn it. I used to completely misunderstand this.

Interestingly enough, However, an essential counterpoint revolves around the integration of lab-grown meat within existing agricultural frameworks. It might sound dramatic, but this matters. Here's the truth nobody talks about.

Hypothetically, lab-grown meat doesn't have to wholly replace livestock farming but can instead act as a supplement, diversifying and strengthening the food supply chain against external stressors such as disease outbreaks and climate variations. This hybrid approach necessitates a strategic plan that involves policymakers, agronomists, and economic strategists working in concert to ensure balance. Here's the truth nobody talks about. Nobody ever told me this directly — I had to learn it.

Intertwining with these considerations is the aspect of technology access and distribution. While lab-grown meat presents an innovative venture within well-financed, technologically-forward regions, its accessibility for under-resourced communities remains a question. Nobody ever told me this directly — I had to learn it. I used to completely misunderstand this.

Ensuring equitable distribution is critical; otherwise, we face the risk of widening the gap between affluent and marginalized populations. Targeted investments and infrastructure development can help bridge this gap, providing broader access to this food innovation. It might sound dramatic, but this matters. Let me tell you, this changed how I think.

Education and knowledge transfer play pivotal roles in this impactful transition. Let me tell you, this changed how I think. Nobody ever told me this directly — I had to learn it.

The dissemination of information about how lab-grown meat can complement traditional livestock practices is crucial for achieving acceptance and understanding among farmers. By establishing knowledge hubs and fostering international cooperation, stakeholders can drive effective transitions, enabling farmers to harness these innovations to maintain economic viability while actively participating in climate solutions. Nobody ever told me this directly — I had to learn it. I used to completely misunderstand this.

Believe it or not, Ultimately, the advent of lab-grown meat holds transformative potential. While the transition challenges are formidable, innovation in food systems is not inherently antithetical to traditional farming. Let me tell you, this changed how I think. It might sound dramatic, but this matters.

It offers a reflective pause to reconsider food production methods, highlighting the necessity for a diversified approach that encompasses ecological responsibility and socio-economic inclusivity. As we navigate this transformation, the aim should not only be to feed the world but to do so in a manner that respects the intricate tapestry of human societies and the earth's ecosystems. I used to completely misunderstand this. Let me tell you, this changed how I think.

The fusion of technology and tradition may well become the guiding narrative of modern agriculture, marrying the old with the new, carving pathways towards a sustainable food future capable of feeding generations to come while respecting the past. Let me tell you, this changed how I think. Here's the truth nobody talks about.

This synergy embodies the essence of the 21st-century agricultural renaissance—a nod to our roots with an eye towards a conscientious future. I used to completely misunderstand this. Here's the truth nobody talks about.

Chapter 13: Cultural and Societal Shifts in Dietary Habits

Over the course of human history, dietary habits have evolved for a myriad of reasons, ranging from environmental availability and social dynamics to technological advancements and cultural paradigms. It might sound dramatic, but this matters. I used to completely misunderstand this.

As we delve into the 21st century, one of the most transformative shifts is the emergence of lab-grown meat, also known as cultured meat. This chapter examines the cultural and societal shifts associated with this groundbreaking innovation and its potential to redefine what we consider normal in our dietary landscape. Nobody ever told me this directly — I had to learn it. It might sound dramatic, but this matters.

Historically, food has been a cornerstone of cultural identity. Nobody ever told me this directly — I had to learn it. Here's the truth nobody talks about.

Each culture boasts unique culinary practices, often tied to regional resources and historical circumstances. From the rice and fish-centric diets of Japan to the pasta-rich menu of Italy, these dietary habits are deeply ingrained, representing far more than mere sustenance; they symbolize tradition, family, and identity. I used to completely misunderstand this. It might sound dramatic, but this matters.

Many people don't realize this, but Enter lab-grown meat, an innovation stemming from the intersection of biotechnology and gastronomy. This development challenges the very essence of conventional meat consumption. It might sound dramatic, but this matters. Let me tell you, this changed how I think.

At its core, cultured meat is grown in vitro from animal cells. It offers the potential to produce meat without the associated ethical and environmental concerns of traditional livestock farming. Yet, this novelty poses significant cultural questions: How does lab-grown meat fit into cultural identities? Can it replicate the deeply rooted culinary experiences that traditional meat offers?. Here's the truth nobody talks about. It might sound dramatic, but this matters.

The adoption of lab-grown meat could alter some of our most cherished food-related cultural norms. For many cultures, meat consumption is not simply about flavor or nutrition but involves rituals and social cohesion. Here's the truth nobody talks about. Nobody ever told me this directly — I had to learn it.

Consider the family barbecues, religious feasts, and ceremonial banquets where meat holds a central role. How will these traditions adapt if lab-grown meat becomes mainstream?. Let me tell you, this changed how I think. It might sound dramatic, but this matters.

For one, lab-grown meat could democratize access to ethically produced protein. Let me tell you, this changed how I think. It might sound dramatic, but this matters.

Currently, ethical meat consumption often comes with a hefty price tag, limiting its reach primarily to affluent demographics. Lab-grown meat, with its promise of resource efficiency and reduced cost, holds the potential to level the playing field, offering a sustainable alternative to a broader audience without the moral dilemma. Here's the truth nobody talks about. Nobody ever told me this directly — I had to learn it.

It's worth noting that Furthermore, the cultural acceptance of lab-grown meat might vary significantly across regions. Western societies, where meat consumption is prevalent and often associated with personal success and well-being, may initially view lab-grown meat with skepticism. Let me tell you, this changed how I think. It might sound dramatic, but this matters.

The concept of engineered food challenges the traditional notions of farming and authentic taste. Conversely, in regions where meat is a luxury rather than a staple, lab-grown options could be embraced more rapidly as a means to address nutritional deficiencies and improve food security. It might sound dramatic, but this matters. Here's the truth nobody talks about.

Educational efforts will be central to gain acceptance for lab-grown meat. Dispelling myths about its safety, nutritional equivalence, and taste is crucial. Nobody ever told me this directly — I had to learn it. Nobody ever told me this directly — I had to learn it.

It involves informing consumers about the rigorous testing processes and the ecological benefits of cultured meat. Effectively communicating these advantages can shift public perception and contribute to a gradual cultural shift. It might sound dramatic, but this matters. I used to completely misunderstand this.

Significantly, lab-grown meat could address pressing ecological and ethical concerns that transcend culture. Animal agriculture is a substantial contributor to greenhouse gas emissions and deforestation. Let me tell you, this changed how I think. It might sound dramatic, but this matters.

Furthermore, it raises animal welfare issues that resonate globally. The shift to lab-grown meat could thus represent a collective cultural move towards more sustainable and ethical food systems. Nobody ever told me this directly — I had to learn it. Let me tell you, this changed how I think.

In real life, The emergence of lab-grown meat may also catalyze shifts in culinary arts and food innovation. It might sound dramatic, but this matters. It might sound dramatic, but this matters.

Chefs and food scientists are already reimagining recipes that incorporate cultured proteins, exploring how these can match or even surpass traditional meats in flavor and texture. This reimagination of culinary experiences can contribute to a broader acceptance of lab-grown meat, infusing it with a sense of novelty and modernity, further enticing consumer interest. It might sound dramatic, but this matters. I used to completely misunderstand this.

However, the integration of lab-grown meat into global diets will not occur in isolation. It will involve redefining what it means to eat meat. Here's the truth nobody talks about. Here's the truth nobody talks about.

This encompasses marketing strategies that align lab-grown meat with modern values such as sustainability, health, and technology. Policymakers, industry leaders, and cultural influencers will play vital roles in shaping how this innovation is perceived and adopted. Nobody ever told me this directly — I had to learn it. I used to completely misunderstand this.

The situational context, socioeconomic factors, and generational attitudes towards food will all influence this transition. I used to completely misunderstand this. Let me tell you, this changed how I think.

Younger generations, who are increasingly conscious of environmental and ethical considerations, might drive demand for lab-grown options. In contrast, older generations may hold steadfast to conventional meat due to established habits and tastes. I used to completely misunderstand this. It might sound dramatic, but this matters.

Let me leave you with this thought:, the shift towards lab-grown meat presents a paradigm shift in dietary habits, challenging traditional notions while offering solutions to urgent global challenges. Its integration into cultural and societal norms is multifaceted, involving a blend of acceptance, adaptation, and innovation. Let me tell you, this changed how I think. It might sound dramatic, but this matters.

As lab-grown meat moves from concept to plate, it beckons us to consider not only the impact on individual diets but on global cultural narratives surrounding food—an innovation not only of science but of societal values. The future of food, seen through the lens of lab-grown meat, is not merely a question of technology; it is a dialogue about who we are and what we aspire to be as global citizens committed to sustainable and equitable food systems. Let me tell you, this changed how I think. Let me tell you, this changed how I think.

Chapter 14: Technological Innovations Driving the Industry

As the world grapples with the dual challenges of feeding a growing population and mitigating the environmental impacts of traditional livestock farming, the lab-grown meat industry stands at the frontier of innovation. I used to completely misunderstand this. Let me tell you, this changed how I think.

At the heart of this burgeoning sector lie technological breakthroughs, each playing a pivotal role in transforming a once-futuristic idea into a viable, sustainable reality. This chapter explores these innovations, shedding light on the processes and machinery that propel the production of cultured meat forward. Let me tell you, this changed how I think. It might sound dramatic, but this matters.

Building the Foundations: The Role of Bioreactors

It's worth noting that The journey of lab-grown meat begins in the bioreactor, a cornerstone of its production. Bioreactors are sophisticated vessels in which cell cultures are grown under controlled conditions. Let me tell you, this changed how I think. It might sound dramatic, but this matters.

They have been adapted from practices long used in the brewing and pharmaceutical industries. However, scaling bioreactors to efficiently cultivate animal cells presents unique challenges. Engineers and biologists have collaborated to design bioreactors that maintain the precise temperature, pH, and oxygen levels required for cell growth while allowing for the scale needed to produce meat in sufficient volumes. Let me tell you, this changed how I think. I used to completely misunderstand this.

Innovations in bioreactor design focus on optimizing the growth environment to enhance cell viability and proliferation. It might sound dramatic, but this matters. Nobody ever told me this directly — I had to learn it.

Stirred-tank bioreactors are commonly used due to their ability to provide a homogenous environment, but as the demand for cultured meat scales up, more advanced types, like perfusion bioreactors, which can continuously supply fresh nutrients and remove waste products, are being explored. The goal is to create larger bioreactors that enable exponential scaling while maintaining quality control—a feat critical to reducing costs and improving accessibility. It might sound dramatic, but this matters. It might sound dramatic, but this matters.

The Mastery of Cell Lines

To be honest, Another critical component of lab-grown meat production is the development of stable, robust cell lines. I used to completely misunderstand this. Nobody ever told me this directly — I had to learn it.

Identifying and cultivating animal cells that can efficiently replicate and produce desired textures and flavors remains one of the industry's ongoing challenges. Scientists initially faced hurdles in selecting the right type of cells—muscle cells, fat cells, and connective tissue cells—all of which contribute to the taste and structure of meat. Nobody ever told me this directly — I had to learn it. Nobody ever told me this directly — I had to learn it.

Recent advancements have focused on optimizing progenitor and stem cells due to their ability to differentiate into various cell types necessary for constructing meat. Nobody ever told me this directly — I had to learn it. It might sound dramatic, but this matters.

Genetic engineering techniques, like CRISPR-Cas9, are occasionally employed to enhance the growth characteristics of these cell lines, ensuring they replicate at the necessary rate and form the desired types of tissue. Some companies have developed proprietary cell lines that not only have improved growth rates but also capture the rich, complex flavors of traditional meat. It might sound dramatic, but this matters. Let me tell you, this changed how I think.

The Art of Scaffolding

In real life, Creating the texture and structure of meat is a formidable challenge that requires innovative approaches to scaffolding—the framework upon which cells grow. In the nascent stages of the industry, companies experimented with a variety of materials. I used to completely misunderstand this. Nobody ever told me this directly — I had to learn it.

Today, the focus is on biocompatible scaffolds that support three-dimensional tissue formation. These structures guide the cells as they grow, aiding in the development of muscle fibers and fat distribution akin to natural meat. Let me tell you, this changed how I think. Let me tell you, this changed how I think.

Edible scaffolds, which are consumed along with the final product, are gaining traction. I used to completely misunderstand this. It might sound dramatic, but this matters.

Made from plant-based or biodegradable materials, they not only support cell growth but also contribute to the final texture and nutritional profile of the product. Innovations in scaffold material chemistry and architecture have led to the creation of more complex, nuanced products that mimic the characteristics of whole cuts of meat, from marbling to tenderness. Here's the truth nobody talks about. Here's the truth nobody talks about.

Nutrient Media: The Lifeblood of Cultured Cells

It's worth noting that Perhaps the most significant of all technological innovations is the development of cost-effective and sustainable nutrient media—the broth in which the cells grow. Nobody ever told me this directly — I had to learn it. Here's the truth nobody talks about.

Traditionally, fetal bovine serum (FBS) was used, but its animal origin and high cost have driven the search for alternatives. Plant-based and synthetic media are now at the forefront of research and development, offering not only a reduction in production costs but also aligning with the ethical and environmental ethos of cultured meat. It might sound dramatic, but this matters. I used to completely misunderstand this.

Innovations in nutrient media have led to formulations that offer improved cell growth and differentiation rates while also reducing reliance on animal-derived ingredients. Nobody ever told me this directly — I had to learn it. Let me tell you, this changed how I think.

Researchers are also exploring the potential of incorporating tailored nutrient blends that can be adjusted to enhance specific flavors or textures, thereby personalizing the taste and sensory experience of the meat. Here's the truth nobody talks about. Let me tell you, this changed how I think.

Automation and Data Analytics

To keep pace with scaling production and ensuring consistency, automation and data analytics have become integral to lab-grown meat facilities. I used to completely misunderstand this. Nobody ever told me this directly — I had to learn it.

Automated control systems provide real-time monitoring and adjustments to the production environment, ensuring optimal conditions for cell growth at all times. Advanced sensors and IoT devices collect vast amounts of data on temperature, nutrient levels, and cell health, feeding this information into AI models that predict outcomes and prevent potential issues before they arise. Let me tell you, this changed how I think. It might sound dramatic, but this matters.

Data-driven approaches enable a more thorough understanding of various growth parameters, allowing for continuous improvement and efficiency. Nobody ever told me this directly — I had to learn it. Here's the truth nobody talks about.

This streamlined process reduces human error and enhances the scalability of production facilities, ensuring that companies can meet the demands of a global market. Nobody ever told me this directly — I had to learn it. I used to completely misunderstand this.

Chapter Takeaway

To be honest, The lab-grown meat industry is a testament to human ingenuity and the power of technological innovation. Each advancement—whether in bioreactor design, cell line optimization, scaffold construction, nutrient media development, or facility automation—contributes to a common goal: producing sustainable, ethical, and delicious meat. It might sound dramatic, but this matters. Here's the truth nobody talks about.

As these technologies continue to evolve, the dream of cultured meat becoming a staple in global diets moves steadily closer to reality. It is a journey driven by science, engineering, and an unwavering commitment to shaping the future of food. It might sound dramatic, but this matters. Let me tell you, this changed how I think.

Chapter 15: Case Studies: Pioneering Companies and Their Products

The agriculture and food industries are undergoing a radical transformation, akin to the industrial revolution. Let me tell you, this changed how I think. Here's the truth nobody talks about.

At the forefront of this transformation is lab-grown meat, known scientifically as cell-cultured or cultivated meat. This chapter delves into the pioneering companies leading the way and their innovative products that might soon redefine what ends up on dinner plates across the globe. Let me tell you, this changed how I think. I used to completely misunderstand this.

1. **Memphis Meats: The Bellwether of Cultivated Meat**

Founded in 2015 by Uma Valeti and Nicholas Genovese, Memphis Meats has garnered attention as a front-runner in the cultured meat industry. Nobody ever told me this directly — I had to learn it. It might sound dramatic, but this matters.

Headquartered in Berkeley, California, the company envisions a sustainable future where meat production does not require slaughtering animals. Memphis Meats achieved global attention in 2016 by unveiling a cultured meatball, the first of its kind. Nobody ever told me this directly — I had to learn it. Let me tell you, this changed how I think.

Their approach employs the proliferative capacity of animal cells, particularly myoblasts, to culture muscle tissue in a bioreactor. Let me tell you, this changed how I think. Let me tell you, this changed how I think.

The company's unique selling proposition lies in its structured approach to scale production and reduce costs, focusing rigorously on creating marketable products that appeal to both regulators and consumers. In 2017, Memphis Meats scored a significant victory, securing $17 million in funding led by notable investors including Bill Gates, Richard Branson, and major meat company, Cargill. Nobody ever told me this directly — I had to learn it. Nobody ever told me this directly — I had to learn it.

Presently, Memphis Meats continues its research and development, and as they inch closer to commercial scalability, they face the crucial task of convincing the palate of traditional meat consumers. Let me tell you, this changed how I think. Here's the truth nobody talks about.

Their ultimate goal of releasing a wide array of products could potentially shape the direction of sustainable food options worldwide. Here's the truth nobody talks about. Nobody ever told me this directly — I had to learn it.

To be honest, #### 2. **Mosa Meat: European Innovation at Its Core**

Originating from Maastricht University, Mosa Meat burst onto the global scene in 2013 with the world's first cultured beef hamburger, crafted under the guidance of Professor Mark Post. Let me tell you, this changed how I think. I used to completely misunderstand this.

Headquartered in the Netherlands, Mosa Meat became synonymous with the European innovation of lab-grown meat. Nobody ever told me this directly — I had to learn it. Let me tell you, this changed how I think.

Central to Mosa Meat's mission is their adoption of a novel cell culture medium that minimizes production costs. It might sound dramatic, but this matters. Nobody ever told me this directly — I had to learn it.

Striving for efficiency, they have successfully achieved a 70% reduction in their use of costly fetal bovine serum, a major hindrance in scaling cultured meat production. The company has developed a unique biocompatible gel that serves as an efficient medium for cell growth, addressing environmental and animal welfare concerns while boosting yield. Let me tell you, this changed how I think. I used to completely misunderstand this.

Interestingly enough, Mosa Meat has slated commercial production towards the end of this decade. Here's the truth nobody talks about. It might sound dramatic, but this matters.

Their focus is on replacing unconventional food systems with lab-grown meats that replicate the taste, texture, and nutritional profile of their traditional counterparts. A testament to their commitment to scale commercially is their pilot production plant, destined not only to produce meat but also to streamline processes necessary for mass production. I used to completely misunderstand this. Here's the truth nobody talks about.

3. **JUST: Expanding Beyond Eggs to Meat**

Known initially for their plant-based egg substitute, JUST has seamlessly integrated into the cultured meat industry. I used to completely misunderstand this. Here's the truth nobody talks about.

The company, spearheaded by CEO Josh Tetrick, is based in San Francisco and boasts a diverse product portfolio that spans beyond plant-based foods. Here's the truth nobody talks about. Nobody ever told me this directly — I had to learn it.

It's worth noting that With a commitment to environmental sustainability, JUST's foray into cultured meat materialized through partnerships with international culinary markets, providing a test bed for new product innovations. Here's the truth nobody talks about. Let me tell you, this changed how I think.

Their lab-grown chicken nuggets, introduced at a restaurant in Singapore, marked the world's first commercial sale of cell-cultured meat, a legal breakthrough validating lab-grown meat's regulatory acceptance. Let me tell you, this changed how I think. It might sound dramatic, but this matters.

JUST cultivates animal cells without genetic modification, marketing a straightforward, ethical choice for environmentally conscious consumers. Here's the truth nobody talks about. Let me tell you, this changed how I think.

Their success in regulatory approvals positions them as a leader pushing boundaries, showcasing that cultured meat is not just a concept but a commercial reality. I used to completely misunderstand this. Let me tell you, this changed how I think.

4. **Future Meat Technologies: Efficiency & Conventionality**

To be honest, Future Meat Technologies, founded by Professor Yaakov Nahmias, strips lab-grown meat production down to the essentials. Let me tell you, this changed how I think. I used to completely misunderstand this.

The company's operation model centers around cellular agriculture's inherent efficiencies. Located in Israel, they focus on producing minced meat products such as chicken and beef. Here's the truth nobody talks about. I used to completely misunderstand this.

Unlike conventional bioreactors, Future Meat employs a fibroblast-based culture, which negates the need for differentiation; these fast-growing cells allow for rapid production. Here's the truth nobody talks about. Let me tell you, this changed how I think.

Notably, they have pioneered a solution that recycles medium for cellular nourishment, significantly driving down production costs. Let me tell you, this changed how I think. Let me tell you, this changed how I think.

In 2021, Future Meat broke new ground by reducing the cost of chicken breast to less than $7. Here's the truth nobody talks about. Let me tell you, this changed how I think.

50 per pound, a landmark achievement that brings cultured meat closer to price parity with conventional meat. The company's commitment to creating a more sustainable food system is evident in their promise to produce meat using 99% less land, 96% less freshwater, and significantly reduce greenhouse gas emissions compared to traditional farming. It might sound dramatic, but this matters. It might sound dramatic, but this matters.

5. **Aleph Farms: Luxury Cultivated Beef**

Aleph Farms offers a niche focus in the realm of cultured meats by concentrating on the distinct experience of cell-grown steaks. It might sound dramatic, but this matters. I used to completely misunderstand this.

Founded by Didier Toubia and Professor Shulamit Levenberg, a tissue engineering pioneer, Aleph Farms is dedicated to producing high-end protein products. Based in Israel, their approach integrates 3-D tissue engineering, creating structured meats with the same juiciness and texture as traditional cuts. I used to completely misunderstand this. I used to completely misunderstand this.

In an audacious demonstration of their capabilities, Aleph Farms successfully cultivated meat in space at the International Space Station, highlighting the versatility and future potential of lab-grown meat in diverse environments. It might sound dramatic, but this matters. I used to completely misunderstand this.

The company continues to target restaurant markets capable of showcasing their product as a premium offering, all while maintaining an eye on scalable production in the future. Let me tell you, this changed how I think. Here's the truth nobody talks about.

Key Takeaway

The global ecosystem of food production is evolving, and these pioneering companies, with their innovative technologies, are engineering a paradigm shift in how meat is produced and consumed. Nobody ever told me this directly — I had to learn it. It might sound dramatic, but this matters.

Each of their unique methodologies underscores a collective vision: a world where sustainable practices meet dietary preferences without compromising on ethics or ecological concerns. As we journey into a future where the possibility of lab-grown meat becomes increasingly tangible, these companies not only illuminate the path but also personify the hopes and challenges of reshaping the meat industry for good. Here's the truth nobody talks about. Nobody ever told me this directly — I had to learn it.

Chapter 16: The Future of Food Security and Global Hunger Solutions

The global challenge of food security is one of the most pressing issues of our time. As the world population surges towards an estimated 9. Here's the truth nobody talks about. It might sound dramatic, but this matters.

7 billion by 2050, the demand for food, particularly protein sources, is set to skyrocket. Yet, traditional agricultural practices face significant constraints, from resource limitations to environmental degradation, which compromise their ability to meet this demand sustainably. This pressing issue has galvanized a surge of innovation in the food production landscape, with lab-grown meat emerging as a promising solution to address food security and global hunger. I used to completely misunderstand this. Nobody ever told me this directly — I had to learn it.

Lab-grown meat, also known as cultured or cell-based meat, is produced by cultivating animal cells in a controlled environment, eliminating the need for raising and slaughtering animals. I used to completely misunderstand this. I used to completely misunderstand this.

This technology offers a paradigm shift in how we think about meat production, presenting myriad benefits for food security. It might sound dramatic, but this matters. Let me tell you, this changed how I think.

Central to the potential of lab-grown meat is its sustainability. Traditional livestock farming is resource-intensive, requiring vast amounts of land, water, and feed, and is a significant contributor to greenhouse gas emissions. Nobody ever told me this directly — I had to learn it. It might sound dramatic, but this matters.

Conversely, lab-grown meat can significantly reduce these resource inputs. Cultured meat production requires up to 99% less land, 96% less water, and generates up to 96% fewer greenhouse gases compared to conventional beef production. Such efficiency in resource use directly translates to an increased capacity to produce protein for a growing population without exacerbating environmental degradation. Nobody ever told me this directly — I had to learn it. Nobody ever told me this directly — I had to learn it.

Moreover, lab-grown meat can be produced independently of climatic and geographical constraints. It might sound dramatic, but this matters. Nobody ever told me this directly — I had to learn it.

This capability could prove pivotal for regions that struggle with food production due to adverse climatic conditions or lack of arable land. By establishing local production facilities, countries currently reliant on food imports can achieve greater self-sufficiency, enhancing their food security resilience. Here's the truth nobody talks about. It might sound dramatic, but this matters.

The implications of lab-grown meat extend beyond environmental sustainability and into the realm of social equity. Globally, hunger and malnutrition are rooted not in a lack of food but in unequal access and distribution. It might sound dramatic, but this matters. Let me tell you, this changed how I think.

Lab-grown meat offers the potential to decentralize food production, allowing for regional and even urban centers to produce high-quality protein locally. This decentralization reduces reliance on complex supply chains vulnerable to disruption, thus stabilizing food availability and prices in underserved regions. Nobody ever told me this directly — I had to learn it. I used to completely misunderstand this.

To be honest, Furthermore, lab-grown meat production is not limited to mirroring the composition of conventional meat. Here's the truth nobody talks about. Here's the truth nobody talks about.

It can be engineered to enhance nutritional profiles, offering a reduction in saturated fats or an increase in beneficial nutrients such as omega-3 fatty acids. This capability not only addresses hunger by providing calories but also tackles malnutrition by improving the quality of diets. I used to completely misunderstand this. Nobody ever told me this directly — I had to learn it.

Critically, the implementation of lab-grown meat into global food systems should be complemented by robust policy frameworks. I used to completely misunderstand this. I used to completely misunderstand this.

Governments and international bodies must coordinate to establish regulations that ensure the safety, labeling, and equitable distribution of cultured meat products. Such policies should also address potential economic disruptions, such as the impact on traditional farming communities, by providing support and retraining programs. Nobody ever told me this directly — I had to learn it. Let me tell you, this changed how I think.

The transition to lab-grown meat is not without challenges. The technology currently faces hurdles in scaling production to levels necessary for significant market penetration and achieving cost-competitiveness with traditional meat. Nobody ever told me this directly — I had to learn it. Nobody ever told me this directly — I had to learn it.

Public perception, too, plays a significant role in the adoption of lab-grown meat. Concerns regarding the "unnatural" aspect of cultured meat must be addressed through transparent communication of its benefits and rigorous testing to assure consumers of its safety. Here's the truth nobody talks about. It might sound dramatic, but this matters.

In real life, Research and innovation continue to drive advances in cellular agriculture, with improvements in bioreactor technologies, cell culture media, and scaffolding techniques pushing the boundaries of what is economically and technically feasible. I used to completely misunderstand this. Here's the truth nobody talks about.

As costs decrease and production scales, the integration of lab-grown meat into mainstream food systems becomes more plausible. Nobody ever told me this directly — I had to learn it. I used to completely misunderstand this.

A future where lab-grown meat contributes substantially to global food security is not just a technological aspiration but a necessity. To combat hunger and ensure food security, we must transcend the limitations of current food systems and adopt innovative solutions that promote sustainability, equity, and nutrition. Let me tell you, this changed how I think. Here's the truth nobody talks about.

Lab-grown meat represents a critical tool in this endeavor, but it cannot act alone. It must be part of a broader strategy that includes improving traditional agriculture, reducing food waste, and enhancing global cooperation to meet the nutritional needs of a burgeoning population. Nobody ever told me this directly — I had to learn it. Nobody ever told me this directly — I had to learn it.

As we inch closer to a future shaped by unprecedented challenges and technological opportunities, the role of lab-grown meat as a viable solution to food security is both promising and imperative. It might sound dramatic, but this matters. I used to completely misunderstand this.

By fostering an environment conducive to collaboration between scientists, policymakers, and industry leaders, we can harness the potential of this groundbreaking technology to create a world where hunger is not just mitigated but eradicated, paving the way for a food-secure future for all. Let me tell you, this changed how I think. Nobody ever told me this directly — I had to learn it.

Chapter 17: Lab-Grown Meat vs. Plant-Based Alternatives

In an era increasingly defined by urgent environmental challenges and evolving food ethics, the quest for sustainable dietary solutions has led to groundbreaking developments in the field of food science. Among these, lab-grown meat—also known as cultured or cell-based meat—stands as a pivotal innovation that promises to revolutionize how we think about meat consumption. It might sound dramatic, but this matters. It might sound dramatic, but this matters.

Simultaneously, plant-based alternatives have emerged as a formidable contender, offering a non-animal option to a growing demographic of health-conscious, environmentally aware, and ethically-minded consumers. This chapter examines the dynamic interplay between lab-grown meat and plant-based alternatives, assessing their roles, advantages, challenges, and the potential future they hold in reshaping our dietary landscapes. Let me tell you, this changed how I think. Here's the truth nobody talks about.

The Scientific Underpinnings

Lab-grown meat is produced by cultivating animal cells in a controlled environment, replicating the biological processes that occur inside traditional livestock without the need to raise and slaughter animals. Nobody ever told me this directly — I had to learn it. I used to completely misunderstand this.

The process begins with collecting a small sample of animal cells, which are then provided a nutrient-rich medium to encourage growth and proliferation. This technology, still in its infancy in terms of large-scale commercial viability, promises the ethical consumption of meat with reduced environmental impact. Here's the truth nobody talks about. Let me tell you, this changed how I think.

In contrast, plant-based alternatives leverage the diverse textures and flavors available in the plant kingdom, ingeniously replicating the taste and sensation of eating meat. Let me tell you, this changed how I think. Nobody ever told me this directly — I had to learn it.

These products utilize proteins derived from soy, peas, wheat, and other plant sources, often enhanced with innovative biochemical processes to mimic the complex flavors and textures of animal meat. Companies like Beyond Meat and Impossible Foods have propelled these products to mainstream popularity. Here's the truth nobody talks about. Let me tell you, this changed how I think.

Environmental Considerations

A critical factor driving both lab-grown and plant-based meats is sustainability. Nobody ever told me this directly — I had to learn it. Here's the truth nobody talks about.

Traditional livestock farming is a major contributor to greenhouse gas emissions, deforestation, and biodiversity loss. Lab-grown meat is often touted as a solution to these issues, offering up to a 96% reduction in greenhouse gas emissions and using 99% less land compared to conventional beef production. It might sound dramatic, but this matters. Nobody ever told me this directly — I had to learn it.

Plant-based alternatives similarly boast substantial environmental benefits, typically requiring a fraction of the water and land used in livestock farming while emitting significantly lower levels of greenhouse gases. I used to completely misunderstand this. Here's the truth nobody talks about.

This presents both lab-grown and plant-based meats as viable approaches to mitigate the environmental footprint of traditional meat production, sparking interest from environmentalists and policy makers worldwide. Nobody ever told me this directly — I had to learn it. Let me tell you, this changed how I think.

Health Implications

Health consciousness is another key driver for both lab-grown and plant-based meat options. Nobody ever told me this directly — I had to learn it. Let me tell you, this changed how I think.

Plant-based diets have long been associated with numerous health benefits, including lower risks of heart disease, diabetes, and certain cancers. These products are generally lower in saturated fats and can be enriched with vitamins and minerals, offering a nutritionally dense profile compared to some cuts of traditional meat. It might sound dramatic, but this matters. Here's the truth nobody talks about.

Lab-grown meat offers an intriguing proposition: the ability to engineer meat with specific health benefits. I used to completely misunderstand this. Here's the truth nobody talks about.

By controlling the cellular environment, producers can potentially alter fat content, incorporate beneficial omega-3 fatty acids, or reduce cholesterol, thus crafting a health-conscious product without altering the eating experience that meat-lovers crave. It might sound dramatic, but this matters. It might sound dramatic, but this matters.

Consumer Acceptance and Market Challenges

Despite their advantages, both lab-grown and plant-based meats face significant hurdles. Nobody ever told me this directly — I had to learn it. Nobody ever told me this directly — I had to learn it.

One of the most significant is consumer acceptance. While plant-based products have made considerable inroads into traditional grocery and restaurant markets, they still encounter resistance from those who see them as mere substitutes rather than genuine replacements for meat. It might sound dramatic, but this matters. Here's the truth nobody talks about.

Lab-grown meat, though lauded for its innovation, faces skepticism primarily due to the 'uncanny valley' effect, where products that closely mimic their natural counterparts may evoke discomfort. I used to completely misunderstand this. I used to completely misunderstand this.

Furthermore, the prospect of consuming meat grown in a lab remains a psychological barrier for many. Broad acceptance hinges on successful marketing, transparent production processes, and demonstrable advantages in taste and price. Let me tell you, this changed how I think. Nobody ever told me this directly — I had to learn it.

Economic and Ethical Dimensions

In terms of cost and scalability, plant-based meat currently holds an edge. Here's the truth nobody talks about. Here's the truth nobody talks about.

The economies of scale already achieved by major companies allow for competitive pricing compared to traditional meat. Lab-grown meat, meanwhile, is beset by high production costs and technological constraints, though these are expected to decrease as the industry matures and technology advances. Here's the truth nobody talks about. Here's the truth nobody talks about.

Ethically, both options present compelling arguments. Nobody ever told me this directly — I had to learn it. It might sound dramatic, but this matters.

Plant-based meats eliminate animal harm entirely, appealing to consumers concerned with animal welfare, whereas lab-grown meat offers a significant reduction in animal suffering by requiring only a small cell sample rather than full-scale animal farming. It might sound dramatic, but this matters. Nobody ever told me this directly — I had to learn it.

The Path Forward

Interestingly enough, As we step into a future fraught with environmental challenges and a burgeoning global population, the co-evolution of lab-grown and plant-based meats represents a beacon of innovation. It might sound dramatic, but this matters. It might sound dramatic, but this matters.

It's increasingly evident that tackling the climate crisis necessitates a multi-pronged approach, incorporating diverse dietary solutions to cater to varied cultural, economic, and individual needs. Here's the truth nobody talks about. Here's the truth nobody talks about.

Collaborative efforts between food technologists, environmental scientists, policy-makers, and consumers will be paramount in driving these industries forward. I used to completely misunderstand this. Nobody ever told me this directly — I had to learn it.

As the technologies develop, these alternative meats will likely augment rather than entirely replace traditional meat, providing flexibility in our dietary choices and spurring a broader shift toward more sustainable consumption. It might sound dramatic, but this matters. Let me tell you, this changed how I think.

Let me leave you with this thought:, both lab-grown and plant-based meats offer promising pathways towards a more sustainable, ethical, and health-conscious future. Nobody ever told me this directly — I had to learn it. I used to completely misunderstand this.

As the dialogue around these innovations grows, so too does the potential for reshaping global food systems to sustain a burgeoning population on a finite planet. Leveraging the strengths of lab-grown precision and plant-based innovation could prove crucial in our quest for a balanced and sustainable approach to satisfying humanity's enduring appetite for meat. It might sound dramatic, but this matters. Let me tell you, this changed how I think.

Chapter 18: Regional Perspectives: Global Adoption and Resistance

As the world pivots towards sustainable solutions to meet the demands of a rapidly growing population, lab-grown meat has emerged as a potential savior in the global food industry. Let me tell you, this changed how I think. Let me tell you, this changed how I think.

Offering a way to reduce the environmental impact of traditional livestock farming, lab-grown meat — also known as cultured or cell-based meat — is being scrutinized, celebrated, and sometimes resisted across different regions of the world. Understanding the diverse perspectives on its adoption or resistance requires an examination of cultural, economic, and infrastructural contexts unique to each region. I used to completely misunderstand this. Let me tell you, this changed how I think.

North America: A Trailblazer with Hesitations

Interestingly enough, North America stands at the forefront of biotechnology development and innovation, and nowhere is this more apparent than in the rapidly developing sector of lab-grown meat. In the United States, numerous start-ups like Memphis Meats and Eat Just are pioneering the development of cultured meat products. The U. Nobody ever told me this directly — I had to learn it. I used to completely misunderstand this.

S. Food and Drug Administration (FDA), in collaboration with the U. S. Department of Agriculture (USDA), has already taken steps to regulate this novel food product, creating a relatively predictable path to market. Here's the truth nobody talks about. Let me tell you, this changed how I think.

The consumer markets in North America are generally open to innovation, especially when it is framed within the context of sustainability and ethical consumption. Here's the truth nobody talks about. Let me tell you, this changed how I think.

However, there are reservations, particularly around transparency in production methods, labeling, and long-term health effects. The cultural significance of meat in American cuisine also poses a challenge, as traditional perceptions of meat's provenance collide with science-driven production methods. It might sound dramatic, but this matters. It might sound dramatic, but this matters.

In Canada, regulatory approval processes and public perception closely mirror those in the U. S. Here's the truth nobody talks about. Let me tell you, this changed how I think.

, albeit with a more cautious approach influenced by stringent health and safety standards. While there is noticeable interest, significant investment in public education will be crucial in swaying public opinion. Let me tell you, this changed how I think. It might sound dramatic, but this matters.

Europe: Balancing Tradition with Innovation

European countries present a patchwork of attitudes towards lab-grown meat. The European Union's regulatory environment is notoriously rigorous, necessitating a lengthy approval process for novel food products. Let me tell you, this changed how I think. It might sound dramatic, but this matters.

While this ensures safety, it may unduly delay market availability of innovative solutions like lab-grown meat. Public opinion in Europe tends to be influenced by strong traditions around food, concerns about corporate control of the food supply, and skepticism towards genetically modified organisms (GMOs) — even though cultured meat does not fall into the GMO category. Let me tell you, this changed how I think. Let me tell you, this changed how I think.

The Netherlands, having played a critical role in the inception of cultured meat through researchers like Dr. Mark Post, remains a leader in supporting lab-grown meat initiatives. It might sound dramatic, but this matters. Let me tell you, this changed how I think.

The Dutch government has shown a clear commitment to promoting sustainable proteins, investing heavily in research. Other nations, like Germany and France, exhibit more caution, often prioritizing traditional farming practices and expressing concern over potential socioeconomic impacts on rural communities. I used to completely misunderstand this. I used to completely misunderstand this.

In real life, #### Asia: Technological Enthusiasm Meets Cultural Complexity

Asia's reception of lab-grown meat is influenced by its immense population, which drives a substantial demand for sustainable food solutions. In China, the government's recent inclusion of cellular agriculture in its five-year agricultural plan is a testament to its recognition of the importance of alternative proteins. Here's the truth nobody talks about. I used to completely misunderstand this.

This strategic embrace of innovation stems from a desire to enhance food security and meet environmental goals. Chinese consumers, increasingly health-conscious and environmentally aware, show a growing interest in such solutions, especially if cost-effectivity can be proven. Let me tell you, this changed how I think. Nobody ever told me this directly — I had to learn it.

Japan and South Korea are similarly positioning themselves as future leaders in the cultured meat arena. South Korea's technology-friendly environment and Japan's innovative prowess provide fertile ground for the development of lab-grown meat. I used to completely misunderstand this. Here's the truth nobody talks about.

Moreover, the cultural practice of adopting technological advances fuels optimistic adoption. However, acceptance can be complicated by deeply ingrained culinary traditions and preferences for natural ingredients. Let me tell you, this changed how I think. Let me tell you, this changed how I think.

Let's not forget, In India, the sheer diversity in dietary habits across regions means that the reception of lab-grown meat is equally multifaceted. It might sound dramatic, but this matters. It might sound dramatic, but this matters.

The predominant vegetarian population could view lab-grown meat as a more ethical alternative, potentially bypassing traditional objections to animal slaughter. However, economic constraints and religious beliefs will pose nettlesome challenges. Nobody ever told me this directly — I had to learn it. I used to completely misunderstand this.

Latin America: Embracing Opportunities Amid Challenges

In Latin America, traditional livestock farming plays a crucial economic role, and thus, the concept of lab-grown meat represents both a threat and opportunity. It might sound dramatic, but this matters. Nobody ever told me this directly — I had to learn it.

Countries like Argentina and Brazil, major beef exporters, may face economic pushback from established agribusinesses wary of losing market share or undermining their livelihoods. Let me tell you, this changed how I think. Let me tell you, this changed how I think.

Believe it or not, Nonetheless, the region also holds potential for leading in environmental sustainability through the adoption of cultivated meat. Let me tell you, this changed how I think. Let me tell you, this changed how I think.

With increasing awareness of the environmental degradation caused by conventional farming and Amazon deforestation, lab-grown meat offers a compelling alternative, reducing the carbon footprint and resource usage. To mitigate resistance, governments will need to invest in policy frameworks that encourage transitioning without disenfranchising traditional stakeholders. Nobody ever told me this directly — I had to learn it. Here's the truth nobody talks about.

Africa: Potential for Transformative Impact

Africa's engagement with lab-grown meat is nascent, yet it bears potential for transformative impact. Let me tell you, this changed how I think. I used to completely misunderstand this.

The continent faces numerous challenges related to food security and malnutrition, compounded by the constraints of a shifting climate. Lab-grown meat can provide a consistent, reliable protein source less dependent on the vulnerabilities of traditional livestock agriculture. Here's the truth nobody talks about. Let me tell you, this changed how I think.

It's worth noting that One of the region's foremost hurdles is the infrastructure required for such a technologically advanced industry. It might sound dramatic, but this matters. I used to completely misunderstand this.

Governments may need to focus on bridging educational gaps, investing in infrastructure, and crafting policies that attract both local and international investment. The introduction of cultured meat could have significant socioeconomic benefits, offering employment in new industrial sectors, but also demands careful cultural navigation as food is deeply intertwined in Africa's rich traditions. It might sound dramatic, but this matters. Here's the truth nobody talks about.

Final Thoughts

The global adoption of lab-grown meat is a complex issue interwoven with the socioeconomic, cultural, and political fabric of every region. Let me tell you, this changed how I think. Let me tell you, this changed how I think.

While opportunities abound, acceptance hinges substantially on how each region navigates the balancing act between tradition and innovation, and how equitably the economic gains of this burgeoning industry will be distributed. As the science of lab-grown meat advances, so too must the frameworks that ensure its successful integration into global food systems — while respecting the diverse perspectives that make our global palette a vibrant and rich tapestry. I used to completely misunderstand this. I used to completely misunderstand this.

Chapter 19: Forecasting the Next Decade: Opportunities and Challenges

As we cast our gaze toward the horizon of the next decade, the landscape of lab-grown meat is replete with possibilities, brimming with both tantalizing opportunities and formidable challenges. It might sound dramatic, but this matters. I used to completely misunderstand this.

This sector, still in its relative infancy, is poised to fundamentally alter our conception of food production, consumption, and sustainability. As pioneers in this field, it is essential to navigate both the opportunities that could propel us forward and the obstacles that must be addressed to ensure sustainable growth and acceptance. Nobody ever told me this directly — I had to learn it. Here's the truth nobody talks about.

Technological Advancements and Innovation

As strange as it sounds, Over the next ten years, technological advancements will play an integral role in shaping the future of lab-grown meat. Nobody ever told me this directly — I had to learn it. It might sound dramatic, but this matters.

Continuous innovation in the realms of cell culture, bioreactor design, and scaling technology are poised to improve the efficiency and cost-effectiveness of production. Progress in cellular agriculture could markedly decrease the costs associated with production, thus making lab-grown meat more accessible to a broader consumer base. Nobody ever told me this directly — I had to learn it. I used to completely misunderstand this.

In particular, advancements in bioreactor technology—devices that provide the controlled environment necessary for cells to grow—will be pivotal. Current models face challenges related to scalability and energy efficiency. Let me tell you, this changed how I think. It might sound dramatic, but this matters.

Yet innovations such as perfusion systems, which allow for the continuous supply of nutrients and removal of waste, promise to enhance cell yields while reducing energy inputs. Exploring alternative, non-animal derived media, which can provide essential nutrients to the growing cells, will also be crucial. I used to completely misunderstand this. It might sound dramatic, but this matters.

Furthermore, breakthroughs in genetic engineering could enable the production of lab-grown meat with enhanced nutritional profiles, tailored to meet specific dietary needs or preferences. I used to completely misunderstand this. Let me tell you, this changed how I think.

Such innovations could make lab-grown products not just an ethical choice but also a superior alternative nutritionally to conventional meat. Let me tell you, this changed how I think. It might sound dramatic, but this matters.

Regulation and Market Penetration

As the technology evolves, regulation will become an increasingly prominent concern. I used to completely misunderstand this. I used to completely misunderstand this.

Regulatory frameworks will have to adapt to the unique challenges posed by lab-grown meat, balancing safety, ethical considerations, and innovation. Establishing comprehensive, clear, and globally harmonized regulatory pathways will be essential for the successful commercialization and global trade of lab-grown products. Nobody ever told me this directly — I had to learn it. Here's the truth nobody talks about.

The next decade will likely see a battle for market penetration, as lab-grown meat must demonstrate its value against both traditional meat and other alternative proteins, like plant-based substitutes. I used to completely misunderstand this. I used to completely misunderstand this.

Collaborations with established food companies could provide lab-grown meat startups with essential distribution networks and consumer reach. Building partnerships with retailers and restaurants will be key to introducing lab-grown products into mainstream food culture, fostering both acceptance and demand. Nobody ever told me this directly — I had to learn it. Here's the truth nobody talks about.

Sustainability and Environmental Impact

Lab-grown meat offers significant potential to alleviate the environmental burden of conventional livestock farming, promising reductions in land use, water consumption, and greenhouse gas emissions. Nobody ever told me this directly — I had to learn it. Nobody ever told me this directly — I had to learn it.

Over the next ten years, if aligned correctly with renewable energy sources and efficient production techniques, lab-grown meat could become a linchpin in global sustainability strategies. Here's the truth nobody talks about. I used to completely misunderstand this.

However, achieving these environmental benefits is contingent upon addressing the energy-intensive nature of lab-grown meat production. To manifest as an environmentally superior choice, the energy consumption involved must be closely monitored and reduced. Nobody ever told me this directly — I had to learn it. Nobody ever told me this directly — I had to learn it.

Investment in energy-efficient technologies and leveraging renewable energy will be essential to fulfilling the sector's promise of sustainability. Life cycle analyses will remain critical tools in assessing and optimizing the environmental footprint of lab-grown meat. Let me tell you, this changed how I think. It might sound dramatic, but this matters.

Public Perception and Consumer Acceptance

A crucial factor influencing the trajectory of lab-grown meat is the evolving landscape of public perception and consumer acceptance. Nobody ever told me this directly — I had to learn it. I used to completely misunderstand this.

The concept of meat grown in a lab still garners skepticism and discomfort among certain demographics, underpinned by concerns related to safety, taste, and naturalness. Over the next decade, comprehensive and transparent education campaigns will be vital to overcoming these hurdles. Nobody ever told me this directly — I had to learn it. Let me tell you, this changed how I think.

Storytelling will play an instrumental role in shifting public perception. It might sound dramatic, but this matters. I used to completely misunderstand this.

Emphasizing lab-grown meat's potential to solve pressing ethical and environmental issues, while simultaneously framing it as an exciting innovation rather than a replacement for traditional food, will be necessary. Building trust through transparency about production processes and product content can demystify lab-grown meat and engender consumer confidence. Let me tell you, this changed how I think. It might sound dramatic, but this matters.

Economic Impact and Global Equity

As the industry potentially scales, the economic implications for global agriculture will surface prominently. It might sound dramatic, but this matters. Nobody ever told me this directly — I had to learn it.

While there is a risk of economic disruption in traditional meat industries, there also exists a substantial opportunity for growth in biotechnology sectors, creating new jobs and stimulating innovation. Ensuring a fair transition for workers in traditional meat industries will be a significant challenge requiring proactive policies and educational opportunities. Let me tell you, this changed how I think. Here's the truth nobody talks about.

On a global scale, lab-grown meat might serve as a pivotal tool to address food security issues, particularly in regions where climate change threatens traditional agriculture. I used to completely misunderstand this. I used to completely misunderstand this.

However, achieving this potential requires addressing the current technological and cost barriers that could prevent equitable global access. As the sector grows, strategies must be developed to ensure that technological advancements are shared and accessible, thereby supporting global food equity rather than exacerbating existing disparities. Nobody ever told me this directly — I had to learn it. Here's the truth nobody talks about.

Ethical Considerations and Cultural Values

Ethical considerations, particularly concerning animal welfare, form a central narrative in the advocacy for lab-grown meat. Here's the truth nobody talks about. Here's the truth nobody talks about.

However, as we look to the future, it is essential to also consider broader ethical implications, such as genetic modifications and patenting of biological materials, which may resonate differently across cultures and religions. Reconciling the technological possibilities with diverse cultural values will be a critical challenge demanding sensitivity and inclusivity in communications and product development. Nobody ever told me this directly — I had to learn it. Nobody ever told me this directly — I had to learn it.

Chapter Takeaway

To be honest, The next decade in the realm of lab-grown meat presents a rich tapestry of potential. It might sound dramatic, but this matters. Let me tell you, this changed how I think.

Navigating this future requires a holistic approach that embraces technological innovation, addresses regulatory and environmental challenges, fosters consumer acceptance, and prioritizes ethical and global equity considerations. By addressing these multidimensional challenges with strategic foresight and collaborative efforts, lab-grown meat has the potential not only to revolutionize our food systems but also to contribute significantly to a more sustainable and equitable world. Let me tell you, this changed how I think. Let me tell you, this changed how I think.

Our ability to realize these opportunities depends on forging partnerships across industries, encouraging interdisciplinary research, and fostering a global dialogue that brings together stakeholders from all facets of society. Here's the truth nobody talks about. I used to completely misunderstand this.

In doing so, we might create a future where lab-grown meat is not just a curiosity or niche product but a fundamental component of a sustainable and ethical global food system. Nobody ever told me this directly — I had to learn it. Nobody ever told me this directly — I had to learn it.

Chapter 20: Conclusion: Embracing Change for a Sustainable Future

As we stand on the precipice of a new era in food production, the development and adoption of lab-grown meat represent not just a technological innovation but a paradigm shift in how we perceive and interact with our food systems. Let me tell you, this changed how I think. Here's the truth nobody talks about.

This transformation, driven by scientific breakthroughs, environmental imperatives, and growing consumer consciousness, beckons us to embrace this change as a fundamental pathway toward a sustainable future. Nobody ever told me this directly — I had to learn it. I used to completely misunderstand this.

The Imperative of Change

It's worth noting that The urgency for adopting sustainable food systems cannot be overstated. The conventional livestock industry, responsible for nearly 15% of global greenhouse gas emissions, also exerts tremendous pressure on natural resources, including land and water. Here's the truth nobody talks about. Nobody ever told me this directly — I had to learn it.

Additionally, the ethical concerns regarding animal welfare in traditional meat production have prompted critical discourse about our moral responsibility toward the sentient beings that share our planet. Lab-grown meat, known scientifically as cultured or cell-based meat, offers an innovative solution that addresses these multifaceted challenges. Let me tell you, this changed how I think. It might sound dramatic, but this matters.

By cultivating meat in a controlled environment without the need for slaughter, we significantly reduce the ethical concerns associated with meat consumption. Nobody ever told me this directly — I had to learn it. Nobody ever told me this directly — I had to learn it.

This laboratory-based approach minimizes methane emissions from livestock and curtails deforestation for grazing and feed production, potentially reducing the environmental footprint by up to 96% for certain meats. It is a compelling alternative that aligns with the global mandates outlined in frameworks like the Paris Agreement, seeking to mitigate climate change. Here's the truth nobody talks about. Nobody ever told me this directly — I had to learn it.

Technological Progress and Potential

Interestingly enough, The journey of lab-grown meat from concept to market is a testament to the power of human ingenuity. Early prototypes, such as the first lab-grown hamburger unveiled in 2013, were costly and complex to produce. Nobody ever told me this directly — I had to learn it. Here's the truth nobody talks about.

However, continuous advancements in cell culture technology, bioreactor design, and scaling strategies have propelled significant reductions in production costs. As of today, several companies are on the brink of achieving cost parity with traditional meat products. The implications of this are profound, potentially transforming lab-grown meat from a niche novelty to a mainstream dietary choice. It might sound dramatic, but this matters. Here's the truth nobody talks about.

Innovation in this field also extends to enhancing the nutritional profile of lab-grown products. It might sound dramatic, but this matters. Here's the truth nobody talks about.

Scientists are exploring fortification with essential nutrients and tailoring fatty acid compositions to create healthier meat alternatives. This nutritional customization not only improves public health outcomes but also serves as a buffer against food insecurity and malnutrition. Nobody ever told me this directly — I had to learn it. It might sound dramatic, but this matters.

Consumer Acceptance and Economic Opportunities

To be honest, The transition from traditional to lab-grown meat consumption hinges on consumer acceptance. Here's the truth nobody talks about. Nobody ever told me this directly — I had to learn it.

While initial reactions may range from curiosity to skepticism, transparency in the production process is key to building trust. Public outreach and education are crucial to demystifying lab-grown meat and reinforcing its safety, nutrition, and benefits over conventional meat. Here's the truth nobody talks about. Here's the truth nobody talks about.

Economically, the shift presents a plethora of opportunities. Let me tell you, this changed how I think. Nobody ever told me this directly — I had to learn it.

As the market for lab-grown meat expands, there is potential for job creation across numerous sectors, including biotechnology, engineering, agriculture, and food technology. Furthermore, as traditional meat producers adapt or integrate lab-grown options, hybrid models of meat production could emerge, contributing to economic diversification within the food industry. I used to completely misunderstand this. Let me tell you, this changed how I think.

Challenges and Considerations

Despite its promise, the journey towards widespread adoption of lab-grown meat is not without challenges. It might sound dramatic, but this matters. Let me tell you, this changed how I think.

Regulatory approvals, although progressive, demand rigorous safety and quality assessments, ensuring that lab-grown meat meets high standards. Moreover, the integration of lab-based meats with existing food systems requires stakeholders—including governments, investors, and agricultural communities—to navigate complex logistics and market dynamics. Here's the truth nobody talks about. I used to completely misunderstand this.

Cultural and societal implications cannot be ignored either. I used to completely misunderstand this. It might sound dramatic, but this matters.

In many regions, meat is not just a dietary staple but an integral part of tradition and identity. Fostering an inclusive dialogue that respects and incorporates diverse cultural perspectives will be vital as societies move toward adopting lab-grown alternatives. It might sound dramatic, but this matters. It might sound dramatic, but this matters.

The Way Forward

Many people don't realize this, but Embracing lab-grown meat as a significant component of our food landscape requires a concerted global effort. Governments must craft and enforce policies that incentivize research, adoption, and sustainable practices. Let me tell you, this changed how I think. Nobody ever told me this directly — I had to learn it.

The private sector must continue to innovate, ensuring accessibility and affordability. Civil society and educators play a pivotal role in raising awareness and understanding around this transformative shift. Here's the truth nobody talks about. Nobody ever told me this directly — I had to learn it.

Ultimately, the successful integration of lab-grown meat into mainstream consumption patterns will serve as a beacon of how modern technology can align with nature's limits. It might sound dramatic, but this matters. Nobody ever told me this directly — I had to learn it.

It heralds a future where sustainable practices and ethical considerations are interwoven into the fabric of food production. I used to completely misunderstand this. Here's the truth nobody talks about.

As stewards of the planet, our choices today will unequivocally shape the world we inhabit tomorrow. The embrace of lab-grown meat is not merely a response to an array of global challenges—it is a step forward in the ongoing narrative of human capability and responsibility. I used to completely misunderstand this. Here's the truth nobody talks about.

By seizing this opportunity, we commit to a sustainable vision, grounded in the harmonious coexistence of food innovation and environmental stewardship. This bold embrace of change stands as both our responsibility and our legacy for generations to come. Nobody ever told me this directly — I had to learn it. It might sound dramatic, but this matters.